Re-Reasoning Ethics

Basic Bioethics

Arthur Caplan, editor

A complete list of the books in the Basic Bioethics series appears at the back of this book.

Re-Reasoning Ethics

The Rationality of Deliberation and Judgment in Ethics

Barry Hoffmaster and Cliff Hooker

The MIT Press
Cambridge, Massachusetts
London, England

This book was set in Stone Serif by Jen Jackowitz. Printed and bound in the United States of America.

Library of Congress Cataloging-in-Publication Data

Names: Hoffmaster, C. Barry, author.
Title: Re-reasoning ethics : the rationality of deliberation and judgment in
  ethics / Barry Hoffmaster.
Description: Cambridge, MA : MIT Press, 2018. | Series: Basic bioethics |
  Includes bibliographical references and index.
Identifiers: LCCN 2017038168 | ISBN 9780262037693 (hardcover : alk. paper)
Subjects: LCSH: Ethics. | Bioethics. | Medical ethics. | Judgment (Ethics) |
  Reasoning.
Classification: LCC BJ1031 .H74 2018 | DDC 170/.42--dc23 LC record available at
https://lccn.loc.gov/2017038168

ISBN: 978-0-262-03769-3

10 9 8 7 6 5 4 3 2 1

Contents

Preface

This ethics book is unique. Its content is different and distinctive. It is not about the moral theories that are taught in philosophy courses and written about in philosophy journals. (We use "ethics" and "morality" interchangeably.) Moral theories are comprised of principles, and moral reasoning is taken to consist of applying those principles to the facts of problems to deduce right answers. The descriptions of problems are brief, and the resolutions that are logically derived are necessary, certain, and universal. That theoretical depiction of ethics is attractive, but it is too simple, too spare, and too abstract. Telling the truth, for example, can be a challenge. You adhere to principles of telling the truth and preserving life, but you encounter a situation in which lying is needed to save a life. What is the right thing to do? Abide by the principle that prohibits lying or abide by the principle that prohibits killing? Which is the right moral principle, and why is it the right moral principle? More generally, how is the right moral theory to be determined?

This book is about real moral problems that are complex, contextual, and dynamic, how these demanding problems are framed, examined, and settled, and how those analyses, constructions, and resolutions can be rational. The conception of ethics presented here is unusual because it is empirical and naturalistic. Examples of moral issues are taken from research in anthropology and sociology. But how can such a depiction of ethics be normative? By recognizing what orthodox moral philosophy ignores: moral judgment. Determining which principles, values, and facts are relevant requires judgment; resolving conflicts between principles and values requires judgment; and applying general concepts to particulars requires judgment. Judgment is ubiquitous in our lives, and ethics is no exception. The normativity of ethics inheres in good judgment, and good judgment emanates from well-designed, rational processes of deliberation. This book is about nonformal reason, a more expansive conception of rationality that is different from and independent of the formal reason of logic and mathematics and that suffuses our lives. Re-reasoning ethics is about forming rational processes of deliberation that produce rational judgments. That is the best that finite, fallible human beings can do.

This book is the culmination of a decades-long collaboration between two philosophers in different fields, one who specializes in bioethics (BH) and the other who specializes in philosophy of science (CH). The common enterprise is not surprising, however, because both the rationality of bioethics and the rationality of science is the nonformal reason that fosters good judgment.

The book has two parts. The first part, "Ethics and Reason," develops the unrecognized nonformal reason that instills a different, augmented rationality into ethics. It is presented by example and by showing how the resources of nonformal reason operate in science, a paradigm of rationality. The second part, "Ethics as Design," explains how moral compromise and a liberated, extended, and enriched reflective equilibrium are used in designing ethical decision making and how deliberation, practices, institutions, and policies are designed. The contents of the chapters are summarized below.

Chapter 1. Introduction. Women who are at risk of having a child with a genetic condition were interviewed about how they deliberated about whether to try to get pregnant. This research provides a real, illuminating example of decision making. The women's genetic counselors believed that the women should decide by applying the principle of maximizing expected utility to their situations. The women disavowed that approach. Instead, they imagined scenarios of what it might be like for them and their family to live with an affected child, and they assessed their reactions to those scenarios. For moral philosophy, by contrast, applying principles is the rational way to decide. Beginning with this example, this book will explain how the deliberation these women undertake is quintessentially rational.

The next two chapters show that the logic-based approach of formal reason creates parallel problems in ethics and in science.

Chapter 2. The Difficulties of Applied Ethics. Real cases are used to show that there is much more to ethics than reaching decisions based on deduction from principles and rules. Because the meanings of crucial terms in norms, such as "euthanasia" and "autonomy," are indeterminate, they have to be clarified and specified. When principles and rules conflict, how is the one that prevails to be determined? What makes a problem a moral problem, and what does being a moral problem mean? After nonformal reason is introduced in chapter 4, chapter 5 discusses how it can manage several of the cases presented in chapter 2.

Chapter 3. The Problems Generalized, Diagnosed, and Reframed in a New Account of Reason. The problematic features of moral situations revealed in chapter 2 have exact counterparts in scientific research. Given that, it is not unreasonable to expect them in every domain. Their appearance in science makes it especially clear that the problem is located in the common conception of reason as logic that currently underlies both ethics and science. To prepare the way for examining the nature of an appropriate rational procedure, chapter 3 surveys the received Western tradition of characterizing rationality formally; then it explains why this conception of rationality cannot handle serious problems and why a radically different approach is required. The natural framework for addressing these matters is a philosophical naturalism that recognizes that humans are fundamentally finite and fallible but possess a powerful ability to learn. This start refocuses rationality around intelligent problem solving.

The following two chapters develop a nonformal conception of rationality, first for science (where the issues are more cleanly drawn) and then more broadly.

Chapter 4. The Problems Resolved in a New Account of Reason. Deliberative judgment formation, or simply judgment, is a core human skill that is largely irreplaceable by rational formalisms. Indeed, it is required to create those formalisms. Judgment is rationally learnable and improvable like other skills; and, like all intelligent creatures, humans need abundant practical experience with contextual problem solving for judgment formation to be sufficiently relevant, attentive, perceptive, and judicious to support intelligent problem solving. Equally telling for the role of judgment, even in learning about our norms, is that norms and normative methods have been massively elaborated historically in both science and ethics as experience has been gained, as is characteristic of knowledge and skills that have to be acquired by the human species through learning. Rationality can itself be learned and improved through experience in problem solving, for example, in the way an experienced lawyer powerfully marshals precedents and analogies. The same self-improvement characterizes an increasingly powerful scientific method, the spread of institutional ethical resolution processes, and so on.

In exercising reason, there are four principal resources or means for the improvement of skilled judgment: observation, the use of both formal and informal reasoning procedures, constrained but creative construction, and systematic critical appraisal. These four bundles of processes are utilized both by individuals and, typically more powerfully, by communal groupings such as research teams. The bulk of chapter 4 is devoted to expositions of the four resources of nonformal reason and their strategic deployment in problem solving.

The discussion closes with responses to these questions: (Q1) What kind of account is to be given of the rationality of a problem solution or a decision? Response: that it was formed through a properly designed deliberative process. (Q2) What is the content of an account of rationality? Response: a properly designed deliberative process utilizes each of the four judgment-improving resources, along with the tools of compromise and reflective equilibrium formation, as feasibly as possible. And (Q3) What should be the relationship between an account of rationality and an account of our empirical cognitive capacities? Response: naturalists hold that both the normative and the empirical accounts are fallible and learnable and as such must be mutually adapted in a respectful two-way iterative adjustment. Adaptation of each is driven by identification and avoidance of failure and adoption of success: the empirical account by success in understanding the actual problem-solving processes humans use, both its strengths and its weaknesses, biases and limitations; and the normative account by success in guiding the investigation of problem solving, improving our investigatory capacities in the process.

The next three chapters consolidate this account of nonformal reason by showing how it deals with the cases presented in chapters 1 and 2, adding two distinctive tools for constructing problem resolutions—compromise and (enriched) reflective equilibrium—and illustrating rational deliberation in more examples.

Chapter 5. The Process of Ethical Resolution: Using the Resources of Nonformal Reason. Real examples illustrate how the resources of nonformal reason are used practically. The first is new. It comes from Myra Bluebond-Langner's ethnographic study of children with leukemia. These children desperately wanted to know what was wrong with them, but their parents and the health care staff remained silent. The children used the four resources of nonformal reason to discover that they were dying. And they remained silent. Understanding why presents us with a poignant, magnificent example of intelligence. The resources of nonformal reason then are used to illuminate the plights of Ms. B, Ms. F, K'aila's parents, and Mrs. Smith described in chapter 2.

Chapter 6. Ethics as Design and Its Two Distinctive Methods. The development of nonformal reason is completed with Caroline Whitbeck's application of design in engineering to the design of moral problem solving, supported by the incorporation of two distinctive methods for rationally designing resolutions of practical problems. From Whitbeck's work the notion of ethics as design for flourishing is developed, thereby expanding the scope of ethics to encompass ethical problem solving from the personal to the political. One of the distinctive methods for solving ethical problems constructs a fully engaged moral compromise, which requires that the unmet demands of the conflicting ethical claims that are compromised are acknowledged, and thus retains the full richness of ethical considerations for future deliberation. The other method creates a wide reflective equilibrium that is liberated (rather than confined to inference and formal theory construction), extended (to encompass diverse forms of moral deliberation), and enriched (to encompass creative moral options of all kinds).

Chapter 7. Designing Deliberation. Deliberation occurs whenever intentional judgment formation is required but especially when important decisions rife with uncertainty, complexity, and incommensurability have to be made. The best way to understand deliberation is by examining the operation of nonformal reason in various examples. The examples of chapters 1 and 2 provide instructive exemplars, and more are presented in chapter 7, drawn from social science research. Deliberation can take the form of dramatic rehearsal or a narrative, for example, storytelling by occupational therapists trying to understand the behavior of a young patient. A doctor and a patient can engage in shared deliberation, and doctors and nurses in a neonatal intensive care unit can use institutional deliberation to discuss their disagreements. When participation is properly designed, deliberation can substantially reduce alienation by helping to reunite reason and motivation. Doing that is inherently ethical, and it also is the foundation of the professional ethics of relationships with patients/clients and other professionals.

The final two chapters expand the scope of ethical design for flourishing from the conduct of professionals to the formulation of public policies.

Chapter 8. Designing Practices, Institutions, and Processes. Ethics is embedded in the practices and institutions of society. Three examples illustrate the communal importance of ethical design. One is Anatol Rapoport's classic illumination of fights, games, and debates. Another is an examination of the Prisoner's Dilemma in game theory. And the last is Lon Fuller's account of institutional design, invoking the notion of "polycentric" problems and different forms of governance, especially mediation, in designing solutions.

Chapter 9. Designing Policies. Designing ethical policies is illustrated with two real cases. One is motivated by Guido Calabresi and Philip Bobbitt's book on *Tragic Choices*. A tragic choice is forced when fundamental values conflict and none of the values can be abandoned. Because cadaver kidneys for transplantation are scarce, decisions about how the kidneys will be allocated have to be made. Neither of the two conflicting values, equality and efficiency, can be renounced. The resolution requires cycling, a process driven by the remaining residue of which value currently is being repudiated more extensively. Cycling shifts the priority of the two values back and forth over time. This example also exemplifies the necessity and the importance of moral compromise, which is anathema to formal moral theory. The other example is a brief presentation of Anthony Culyer's illuminating account of how the National Institute for Health and Clinical Excellence (NICE) in England and Wales designed a deliberative process for assessing new health care technologies.

Acknowledgments

Excerpts and/or adaptations from five articles that we have published in journals appear in this book. In chapter 1 an example from a publication in *Bioethics* (Hoffmaster and Hooker 2009) that shows how women decide whether to try to get pregnant after receiving genetic counseling is used to appreciate what real ethical decision making is. In chapters 3 and 4 an extended publication in *Axiomathes* (Hooker 2010) is used to develop a critique of logic-based rationality (chapter 3) and an alternative account of nonformal reason (chapter 4). Chapter 5 presents an article from the *Hastings Center Report* (Hoffmaster 2011) that depicts searches conducted by children with leukemia who want to know what is wrong with them to illustrate the rationality of these non-formal procedures. In chapter 6 an article in *Social Theory and Practice* (Hoffmaster and Hooker 2017) portrays a disagreement between a nurse and a doctor in an intensive care unit to show how a compromise can be constructed, supplemented by an explanation of how moral compromise can be rational. Chapter 9 is about designing policies, and the primary example is the development of a policy for allocating the scarce supply of cadaver kidneys for transplantation, published in an article in the *Milbank Quarterly*[1] (Hoffmaster and Hooker 2013). We are pleased to have had our work appear in these journals, and we are grateful to have this work made available here.

We have been fortunate to be guided by Philip Laughlin, our editor at the MIT Press. He received a novel and creative, but expansive and discursive manuscript, and he spurred the authors to turn it into a tighter, more integrated book. We thank him for his help and his attention. We also thank the three anonymous referees who wrote acute criticisms, suggested additional issues to examine, and recommended extra readings.

We certainly need to thank our families too, especially given how long we have been working on this book. Barry is grateful to Sylvie, his wife, for her understanding of what this project has meant to him, and, even more, for her acceptance of the long time it has taken. He hopes his children, Samantha and Nicolas, might find some of their father in this book. Cliff thanks wife Jean for her unfailing patience, interest, and support and grandchildren Selkie, Ursula, and Arthur for keeping him young at heart and with a well-stocked immune system.

I Ethics and Reason

1   Introduction

What is "re-reasoning ethics"? Re-reasoning ethics is about developing and defending a conception of rationality for ethics that is expansive, flexible, and effectual. It is independent of deductive reasoning, a paradigm of rationality that dominates principle-based moral philosophy. The conclusion of a deduction is necessary, certain, and universal, and those are the features that propagate the universalizability that orthodox moral philosophy demands of ethics. In moral philosophy ethics cannot be contingent, subjective, or relative. If it is wrong for a teacher to criticize a student in front of the class, it is wrong for every teacher. When deduction is taken to be the rationality of ethics, morality becomes a moral theory comprised of moral principles that serve as the major premises of deductive reasoning. A moral theory can have one supreme foundational principle, for example, Kant's categorical imperative or act utilitarianism, or a plurality of principles, for example, intuitionism.

Assimilating ethics to deductive reasoning, principles, and moral theories produces an attenuated rationality and a constricted ethics. This influential and widespread, but misguided, depiction of ethics is the result of condensing rationality to deduction. Problems quickly proliferate in an ethics of applying moral principles. How is a principle justified? By deducing it from a higher principle? If so, how is the regress stopped? How is the relevance of a principle determined? How are the general concepts in principles to be construed? How is the relevance of facts ascertained? How are conflicts between principles resolved? Only after all of these questions are answered could deduction give answers to moral problems by subsuming facts under principles. Moral philosophy has no answers to these questions.

Moral problems have to be identified, framed, analyzed, and assessed before a judgment can be arrived at about what to do. That requires deliberation. Potential ways of resolving problems have to be surveyed and evaluated. Relevant issues are interrelated, so the ways they impinge on one another have to be identified and assessed, and the depictions of the issues have to be constantly examined and adjusted. All the elements of the deliberation and their interactions have to be scrutinized. Is the problem framed properly? What responses could handle the problem? What are the strengths and the weaknesses of the various resolutions? How should the options be compared? How can a decision be made? When should the deliberation stop? These matters, and endless others, require judgment. But how can judgments be rational? They are not deduced. The rationality of judgments inheres in a more expansive conception of rationality.

How can there be rationality beyond deduction? The rationality of deduction resides in the form, not the content, of an argument. For that reason we will call this kind of rationality *formal reason*. If rationality were all and only formal reason, then nothing but logic and mathematics would be rational. There would be little rationality in our lives. Yet we make judgments ceaselessly, often unknowingly. Many are trivial, some are important, a few are momentous. Sometimes we make judgments carelessly or frivolously, but at other times we make judgments carefully and rationally. When we have to make hard decisions, we examine the options extensively, we deliberate seriously, and we make judgments.

We cannot be sure that we made the right decision, but we know that we made a rational decision. A judgment is rational when it emanates from a rational process of deliberation. We will call this kind of rationality *nonformal reason*.

In chapters 3 and 4, the core of this book, nonformal reason is explained and defended by showing that the rationality of science, a field that is regarded as eminently rational, is nonformal reason. Because the orientation of this book is empirical and naturalistic, many real examples are used, not just to elucidate but to vindicate. A real example is presented in this introductory chapter. It comes from research about how women who are at risk of having a child with a genetic condition decide whether to try to get pregnant. The women's genetic counselors believe that their clients should decide by applying the principle of maximize expected utility to their situations, and that doing so would give them the right answer. The women discard that principle. They design a process for deliberating about what is meaningful to them and what they should do. This book explains how their deliberation is rational and how, when deliberation is rational, a judgment that emanates from that deliberation is rational.

To be more explicit, we use "reason" to refer to a conception of the nature and scope of rationality, and use "reasoning" to refer to the specific activities undergone in exercising rationality. We use "formal" and "nonformal" reason to name two different conceptions of rationality, and "formal" and "informal" reasoning to refer to two distinct kinds of activities that may occur under them. In this scheme formal reason contains only formal reasoning, while nonformal reason includes both formal and informal reasoning. We take the reasonings focused on or organized around a particular subject or problem to be deliberations. Deliberations are rational when they are appropriately organized (see chapter 4). All judgments issue from deliberations, however brief or extended, explicit or tacit. Judgments are rational insofar as the deliberations that gave rise to them are rational. Judgment can be improved by learning how to improve deliberation; this includes judgments about improving rational deliberation (see chapter 4).

1.1 Background: Clinical and Philosophical Orthodoxy in Genetic Counseling

The following vignette recounts a meeting among three genetic counselors, Bill Smith, Joseph Giordano, and Albert Samuels, at which Charles Bosk, a medical sociologist studying their work, also is present (Bosk 1992).

> It is a Friday afternoon at postclinic conference. Bill Smith is providing some follow-up on the Meullers, a couple who had recently been in for counseling. Bill explains that the Meullers are a couple who have recently had a child with Von Recklinghausen's disease. Mr. Meuller has four café au lait spots (a marker of this dominant condition in its less serious manifestations). He has been seen by a number of physicians, two of whom thought the spots meant that he carried the gene. Bill then asked Giordano to continue. He first described the counseling session he had had with them and then described a conversation with the Meullers' pediatrician. The pediatrician knew how much Mrs. Meuller did not want an abortion and was angry at the clinic for assigning carrier status to Mr. Meuller without a skin biopsy.
>
> Giordano continued: "To make a long story short, Mr. and Mrs. Meuller went shopping for a dermatologist who said that the spots did not definitely mean that he carried the gene. Mrs. Meuller called today to say that she was going to keep the pregnancy."
>
> Giordano, Smith, and Samuels are all distressed that Mrs. Meuller is going to carry this pregnancy to term. They all believe that there is at least a 50 percent chance that the infant will have the disease (although the seriousness of expression ranges from Mr. Meuller's café au lait spots to the features of the Elephant Man.)
>
> Giordano sums up with an expression of helplessness. Having sized up his options, he says, "The only thing I can accomplish if I intervene is harm."
>
> Bill Smith says, "I want to sum up for your benefit, Bosk. This is a couple who have chosen to minimize and deny the risk, who are bound and determined to

keep the pregnancy. This is a different situation than the Allegras' (other clinic patients), who are trying to decide whether or not to take the risk. These people (the Meullers) are ignoring it entirely. There are three different situations to worry about. Cases where parents accept a very high risk, cases where they deny there is any risk at all, and cases where they can't get the meaning of the risk clear in their own minds." (Bosk 1992: 46–47)

This vignette displays three central features of the way genetic counseling is perceived and conducted.

First, genetic counseling is about risk. But what does "risk" mean here? How are these genetic counselors using the term when they talk about "taking the risk," and the like? It is useful to distinguish two notions of risk. In a narrow sense, risk is simply the probability that a child will be born with a particular genetic disease. In a broad sense, risk is about not just the probability of that bare event, but also includes the manifold ramifications of what it could be like to live with a child who has that genetic disease. Some doctors may consider that only narrow risk is relevant, but it is clearly wide risk that ultimately matters to patients.

Second, uncertainty pervades the determination of both risks. In this vignette the probability of a child having a genetic disease is not clear because it is not known whether Mr. Meuller carries the gene for Von Recklinghausen's disease. That uncertainty is compounded by the widely varying and unpredictable expression of Von Recklinghausen's disease. In its mild form the disease causes areas of hyperpigmentation (café au lait spots) on the skin. Individuals without the disease also can have this hyperpigmentation, usually in milder form. In its extreme form the disease can cause grotesque deformities,[1] mental retardation, epilepsy, skeletal abnormalities, gigantism of a limb, and a variety of tumors. Moreover, for reasons that remain unknown, the disease tends to be more severe when transmitted by a male. So even if Mr. Meuller were a carrier of the disease and if his child were to be born with the disease, how severe the disease would be in that child is unclear.

Third, people who receive genetic counseling are expected to make their subsequent decision on the basis of the (wide) risk they face. The assumption is that when the risk is too high, they (a couple or a woman) should not try to conceive or, if they already have conceived, they should not continue the pregnancy. Giordano, Smith, and Samuels are upset because they believe that the risk for the Meullers is too high and consequently that the Meullers' decision to continue the pregnancy is not justifiable.

Here "risk" is understood in the terms of standard decision theory: First identify a range of narrow risks (probabilities) that describe the relevant range of outcome possibilities and describe these outcomes, then (assuming value is unaffected by probabilities) determine the value of each outcome to the affected party, and multiply it by its narrow risk (probability) to yield (narrow) risk-weighted (or "expected") outcome values (utilities) for each option. The standard decision principle is, then, to choose the option with the highest or maximum expected utility (MEU). In the present case, whether to create or retain a pregnancy should be decided by choosing the outcome that would produce the greater of (a) the expected value (utility) of leading a childless life or of adopting a child (which will have its own wide risks) and (b) the expected value of having a child, but one having the genetic outcome for that option. By tacit agreement this also becomes the morally appropriate decision. The effectiveness of genetic counseling is correlatively assessed in terms of how closely the ensuing decision comports with the result of the MEU process (assuming "typical," or perhaps clinically assessed, utilities for the Meullers). Smith, Giordano, and Samuels evidently are troubled by the decision of the Meullers because they cannot understand how the Meullers' decision maximizes their expected utility. They conclude that the Meullers have failed to grasp and use the proper decision-making procedure. But, strictly, this is beyond their competence.

The standard formulation of the MEU principle incorporates a sharp distinction between facts (the descriptions of wide risks) and values (their evaluation), assigning them separate and independent roles. That tacit theoretical representation of the dichotomy in turn structures the practice of genetic counseling and defines the roles of its practitioners. It is the source of the widely espoused view that genetic counseling must be "nondirective," which limits the responsibility of a genetic counselor to providing only facts, i.e., probabilities and physical consequences. Determining the relevant values and the importance of those values is the sole responsibility of those being counseled, so genetic counselors should in no way try to influence or direct them.

How well does the MEU approach to decision making suit the plights of those who receive genetic counseling? How do they evaluate the wide risks—i.e., probabilities (narrow risks) and their outcomes—explained to them? How, as Smith puts it, do they get the meaning of risks "clear in their own mind"? How do they deliberate and make their difficult decisions?

1.2 The Example: Real Ethical Decision Making after Genetic Counseling

The work of Abby Lippman and F. Clarke Fraser reveals how radically the actual decision-making processes of women at risk of bearing a child with a genetic condition depart from the prescriptive MEU model.[2] We will focus on the women because they were the principals in this research.

How do these women conceive their problem and work their way through it?[3] They begin by minimizing or ignoring the probabilities conveyed to them by genetic counselors, instead formulating their situation in binary terms: "Either we will have an abnormal child or we will not" (Lippman-Hand and Fraser 1979b: 57, 1979d: 332). Even if the denominator in the probability they are given is large, the women focus on the 1 in the numerator and realize that their child could be that one. Two women explain:

> Even if it's a 6% chance ... there will still be that chance of a kid eventually having it.

> I know you doctors have a very good heart, you want to convince us there is always a better chance if you look after this and that. But there is always a chance that it can flip and [will happen]. This scares me. (Lippman-Hand and Fraser 1979a: 120)

The critical factor for these women is a residual, ineliminable possibility of having a child with a genetic condition. Whether the probability is relatively high or relatively low, that prospect cannot be removed—the 1 in the numerator always will be there—and that is what is relevant to them. This feature persists however much the uncertainties are reduced: even if it were possible, knowing the precise probabilities and correlative consequences would not change the necessity of facing up to the possibility of that outcome.[4]

How, then, do the women respond to this invariant feature of their predicament? They construct what Lippman and Fraser call "scenarios"—imagined accounts of what it would be like to have a child with the presaged genetic condition—and assess whether they could cope with the worst scenarios they can envisage. Two factors figure prominently in the way women apprehend their problem and thus in how they formulate and react to their personal scenarios: an appreciation of the amorphousness and complexity of their plight, and a sensitivity to the burden of having to make a decision.

The situations in which these women find themselves are ill-defined, fuzzy, and incredibly complicated. They do not know how serious their child's condition, should it eventuate, would be because of the variable expressivity of genetic diseases. They must consider everyone who is likely to be affected—the child, themselves, their partner, extended family, friends, and society in general—surmising what the plausible, but often uncertain, effects on these people could be—including transformation of their self-images, goals, and values—and what range of responses might be acceptable and what would not, and then appraise those effects. From their own perspectives, they must work through how the multifarious outcomes could impinge upon them in their roles as parent (or potential parent), spouse, relative, and member of various social networks. The sheer number, variety, and complexity of the factors they need to consider are compounded by the shifting interactions among them and the incommensurability of their evaluations.

Moreover, there are no established social norms to structure or guide their decision making. Because the technologies that create these problems are comparatively recent and rapidly developing, because this kind of decision is relatively uncommon, and because public discussion of the issues surrounding such decisions is muted, social guidelines have not evolved and crystallized. Genetic counselors commonly are asked for guidance, along the lines taken by one couple:

> This is a high risk, one in four. Are there others like us? What do people do with this risk? Do they take the chance? What do doctors suggest they do? We've never had to face this kind of thing before, so [we] want to know what others say. (Lippman-Hand and Fraser 1979a: 121)

Such a request can be regarded as a flight from responsibility, as an attempt to relinquish autonomy and avoid the burden of having to decide, and, given how onerous the decision is and how much is at stake, it indeed might be that for some women. Often, however, there is a more charitable interpretation.

Those who make this kind of request are not seeking to be told what to do or soliciting approval of their decision. Rather, they are searching for missing information that can help them frame their problem and give them a sense of what courses of action are reasonable. They are seeking societal direction or standards that can help them reduce their uncertainties and understand their options:

> Once they [the women] have expressed the uncertainties and ambiguities confronting them, their requests for advice may be seen as a search for cues to the range of behavior that might be appropriate in their novel circumstances. In the many spheres of life in which one participates, there are rules and constraints imposed on one's activity by outside forces as well as a variety of role models against which one can judge "what is done" in specific situations. These cues orient the actor in choosing a course of behavior. Parents facing the chance of having an abnormal child do not have such guidelines for their reproductive behavior. Awareness that all who undertake childbearing risk having an abnor-

mal offspring is at best subliminal, and parents are unlikely to know of others in a situation similar to theirs whose behavior might provide some sort of social framework within which they can consider their own options. Thus, these questions may be a search for such guidelines. Even when the question is put more directly—"What would you do?"—it may be part of this search for what would be socially acceptable and reflect the desire to explore the limits of their alternatives. (Lippman-Hand and Fraser 1979a: 124)

What the women desire, in other words, is help with framing and structuring an amorphous, complicated, messy problem and with delimiting the range of acceptable responses. Understood in this way, their requests are an attempt to exercise responsibility, not abdicate responsibility.

Yet as these women create and explore their scenarios, they discover a plurality of moral norms relevant to their plight. The principal norms pertain to not causing pain and suffering and to the multiple responsibilities of parenthood. The latter include the responsibilities of becoming a parent and being a parent to a child, the responsibilities of parents to other members of their families, the responsibilities of parents to society, and the responsibilities of parents to themselves. There are, however, no higher norms that identify which norms, of all the ones a woman might survey, are strongly relevant to this particular problem; that determine the comparative importance of the norms that are relevant; and that resolve conflicts between relevant norms. Nor are there higher norms that direct how relevant norms are to be elaborated and specified as the details of scenarios multiply and mutate or how challenges to the correctness of relevant norms are to be assessed. In addition, there are no methodological norms that govern how to conduct inquiries and deliberations about these matters rationally.

At the same time, the women are acutely aware of the burden of being forced to make this kind of decision. Although there is a general background risk of abnormality for all children who are conceived, women who receive genetic counseling face a heightened risk that renders their decision making more difficult and that exposes them to special scrutiny from their family and friends, from members of the health care professions, and from society. The onus of having to make a choice that most other women do not have to make is itself a significant component of their difficulty. Thus these women not only confront a momentous and seemingly intractable decision, they also have to assume the explicit responsibility that attaches to whatever decision they make.

How do these women work through the scenarios they imagine? Strategies of simplification and neutralization help them to defuse the amorphousness and complexity of their situation, to construct their problem in a manner that makes it manageable. This involves, as we have seen, reducing it to binary terms and trying out worst-case scenarios to see whether they are acceptable. Because these women are highly motivated to have a child, or another child, their problem becomes one of whether they can neutralize the worst outcomes they can imagine. They are searching for a reproductive "green light," and to get that green light, they must feel that they are able to cope with their worst-case scenarios and able to accept responsibility for such a risky undertaking.

Two factors help these women come to grips with their uncertainties. One is the presence of an unaffected child. A woman who already has an unaffected child realizes that she need not "lose," i.e., lose the experience of raising an unaffected child, if she takes a chance, and that reassurance makes it easier to go ahead. The existence of an unaffected child also relieves the pressure on women to "normalize" their biological and social roles as mothers. Finally, insofar as a mother considers that she is acting in the interests of another child who needs a sibling, an unaffected child can help diffuse the feeling of responsibility for a decision to proceed. The other factor is what a woman already has gone through, either as affected by the genetic condition herself or as the mother of an affected child. These experiences concentrate and constrain the imaginative construction of scenarios and provide a touchstone for assessing one's ability to cope with the possibilities that are canvassed. Living with a condition enables a woman to devise more detailed, focused scenarios and to conclude either, "I've been through it and it's not so bad," or "Nobody should have to go through what I've been through." The birth of an affected child, regardless of whether that child lives or dies, strengthens the disposition to see outcomes in binary terms but also provides a basis for estimating whether one could endure a repetition of the experience. Thus these two kinds of experience can annul some of the pervasive uncertainties that confound these women and make it easier for them to decide one way or the other. But only some women have those experiences.

Women who feel they could cope with the most dire problems they can imagine of caring for an affected child, as well as with the responsibility of bringing about the birth of an affected child, try to conceive. Women who feel they could not manage one or both try not to conceive.[5]

The actual decision making by these women is a dramatic departure from applying the MEU principle, but it is not a surprising departure because the prescriptive MEU model does not serve them well. The MEU principle delivers clear decisions only when both the suite of possible outcomes and the probability for each of those outcomes are known and the value of each outcome is clear; in more realistic situations one or more of these elements is often unavailable. For example, genetic counselors can often provide only vague or conditional rates (probabilities) and outcomes, either because it is impossible to make a firm diagnosis, because a diagnosis cannot be associated with a single hereditary pattern, or because a person's genetic status cannot be definitively determined. In those circumstances, the MEU model provides little or no purchase on the problem. More important, even if precise numerical rates and outcomes could be provided, the real issues still would not be addressed. As we have learned, the women generally ignore the probabilities or convert them into yes/no options. They restate and interpret the percentages they have been given to emphasize the possibility that cannot be dispelled—having an affected child—and it is this enduring possibility that is most salient to their decision making.

Nor should the making of clear value assessments be treated as unproblematic, for example, by considering it a private matter for each woman. The decisions these women have to make engage a far wider range of values than just those that concern the condition of the child. Such a broad span of values typically is not represented in the simplistic choice models offered in genetic counseling. From this perspective as well, the MEU decision framework used in counseling proves deeply impoverished and consequently provides the wrong orientation to this decision making.

Nor will expanding the scope of the MEU principle alleviate these problems, for two important reasons. First, genetic counselors are not equipped to provide probabilities for how partners and others involved will respond, or for how women will come to feel about themselves after giving birth, or for an array of other broader consequences. Neither, in general, is anyone else better placed to do so. Second, and more profoundly, many of the values involved are ones that the women themselves have never considered, let alone consciously examined and accepted. Now they have to construct and appraise these values for themselves as their inquiry proceeds. They are forced to consider why they want to have a child. Why is it important to the woman to be a mother, and how genuine is that goal? How much would she be doing this for herself, and how much to please or satisfy others or to fulfill societal expectations? Each woman has to assess other matters that are important to her. How deep and durable is a commitment to a spouse or partner, and how might caring for a child with incessant emotionally and physically draining needs jeopardize that relationship? How much does a job or career mean to her? Would she lose raises in salary and miss promotions because of unremitting demands on her time? How strong are her bonds with her family, and how much help could she reasonably expect from relatives without alienating them? How much does she prize independence and control, and how extensively might she become reliant and dependent on others, including busy health care professionals and strangers? Each is forced to consider the expectations she has for a child. What kind of person would she like her child to be? What achievements would make her proud? How would she react to a child who, in many respects, might not be like most other children? How could she cope with the teasing and taunting that might be inflicted on someone who is particularly vulnerable yet completely innocent and, most important of all, is her child? These women have to scrutinize, honestly and brutally, whether they have the resources to be the mother of a child who has a genetic condition. Do they have sufficient patience, courage, strength, and humility? Can they make the sacrifices that would be required, and what, in the end, would make those sacrifices worthwhile?

It is hard to even identify all the discrete values these questions implicate, let alone claim that those values already exist in a determinate form and are quantifiable in a fashion that would allow them to be multiplied with numerical probabilities. Decisions about motherhood produce the usual hopes and dreams alloyed with the usual worries and fears. Above all, though, these decisions are about love, commitment, and altruism. These matters might not have been contemplated before, certainly not combined as they now must be and in as much depth as they now must be. The prospect of having a child with a genetic condition precipitates a fundamental reappraisal of what a woman's life has been about and what she has regarded as important. To propose that these women should be applying their utilities to probabilities distorts and trivializes what is at stake. They are not plugging values into a formula. These women are deciding what kind of person they want to become, what kind of social matrix they want to live in, and whether they and their families have the endowments and the resources to accomplish that. Rather than applying clear and determinate antecedently existing values, they are creating, modifying, and exploring values. It could be contended that the MEU principle is to be applied only after these value identification and assessment processes are completed, but doing that would be needless because by then it would have become clear whether there is a "green light" to proceed, and whatever finer discriminations MEU might provide would, at best, be superfluous. At worst, trying to apply the MEU principle would prove a diversion from the crucial work of sorting out value issues, and the results could not be trusted at any place where they might suggest a different decision.

The counselors themselves recognize that, even when the numbers are definitive, for example, a 10% chance or a one-in-six risk, their implications are vague. The numbers must be imbued with meaning, and the women do that by incorporating them into overall depictions of their problems:

> In the absence of certain knowledge of whether or not a future child would be affected, the important issue became what having an affected child would mean. Moreover, one's numerical risk only had operational meaning when it was perceived in this context. It was not so much that the parents' perceptions of a diagnosis as severe or otherwise influenced their interpretation of a rate figure, but that this probability was, in itself, seen as just one of the consequences of being "at-risk" (in a general sense) for having an abnormal child: Along with the severity of a condition and its duration, its chance of happening was one of its characteristics. And in incorporating rate information this way parents again seem to be attempting to make it personally meaningful. What could happen, with all its implications and uncertainties, was the issue to be faced. (Lippman-Hand and Fraser 1979a: 121)

As Lippman and Fraser point out, both counselors and those who have to decide recognize that numerical rates have to be interpreted and that "putting a rate 'into perspective' involves closely integrating it with the consequences it entails and the life circumstances of the family involved—something far beyond the simple provision of facts, and a reflection of the need for this kind of information to be processed to become meaningful" (Lippman-Hand and Fraser 1979a: 125).

Lippman and Fraser rightly conclude that this kind of understanding requires a different account of decision making:

> The constant interaction of risk and consequences ... suggests that this is an irreducible complex to be processed in subsequent reproductive decision-making and supports the need for a model of such decision-making that involves the heuristic processing of complex information rather than one that assumes a traditional cost-benefit (risk-burden) approach. The latter may be more amenable to formal decision analysis; the former seems more characteristic of what really happens. (Lippman-Hand and Fraser 1979a: 124)

The MEU principle cannot capture the complex, dynamic, contextual, and holistic nature of the problem these women confront because it reduces the formulation and resolution of the problem to two discrete factors that are defined in simple terms and operate independently. Forced application of the spare, albeit elegant, risk-neutral MEU formula does not just crimp and distort what is at issue, it robs the problem of its deep moral import and its personal momentousness.

This account of how women actually make reproductive decisions after genetic counseling reveals the manifold ways in which their interpretation of information, their framing of problems, and their decision making are contextual. Above all, they are struggling to make the information they receive meaningful by personalizing it to themselves and their situations. To accomplish that, they must put the information in a historical and social context: they must integrate it into their own life histories and the life histories of those close to them. The significance the information assumes will depend, for example, on whether a woman (or someone close to her) is affected by a genetic condition. If so, that experience can give shape and substance to the scenarios that are imagined, and how burdensome that disease has been can influence reactions to those scenarios. The information also must be put in a family context. Whereas a woman who has no children might feel pressure to "normalize" her role as a mother, a woman who already has an unaffected child would not be similarly influenced. On the other hand, a woman who already has a child might feel pressure to provide a sibling with whom that child can play. And the information must be put in a societal context. What resources does society provide to help those who care for a child with a genetic disease? How does society expect these decisions to be made? How do people in similar circumstances decide, and what social norms exist?

These women confront one of the hardest decisions a person can make. Their problem is difficult not just because of all that is at stake, but because it is radically open and indeterminate. Their task is not simply to solve a problem but first to formulate it: to structure their problem and then to devise strategies for approaching and settling it. Moreover, there is a dynamic, reciprocal relationship between attempted formulations of their problem and attempted resolutions of their problem. An effort to crystallize the problem will suggest ways of resolving it, but pursuing those resolutions will ramify back into the reformulation of the problem, and that in turn may require revising the approach. For instance, strategies for dealing with a particular scenario (say, a father not coping) will generate new considerations (what kind of help would the father be?) and insights (extended family and other social resources may be important to outcomes) that lead to modifications and extensions of the scenario and to alternative scenarios (e.g., divorce), and features of these in turn will often suggest new aspects of the problem not yet properly captured and new methods for doing so. Formulation, method, and resolution are interdependent and reciprocal processes. In addition, these women have to find a way of stopping this dynamic exchange. They have to figure out what kind(s) of outcome(s) would be satisfactory and how to recognize them. All this forces the women to construct new conceptions of themselves as persons and to try out those conceptions. They must determine whether a particular conception of what would give their life worth and integrity is possible for them—whether that conception fits their idiosyncratic histories and their manifold potentialities—and whether a particular conception is desirable for them—whether it demands sacrifices that are too onerous and whether it allows them to realize some of their hopes and dreams or at least some of their aspirations.

1.3 Philosophical Implications

Women who receive genetic counseling struggle with what Plato and Aristotle recognized as the fundamental moral question: how ought one to live? Now, however, the subject matter of ethics is different. Edmund Pincoffs regards the prevailing conception of ethics as "so general that it would be tedious to document it":

> It is that the business of ethics is with "problems," that is, situations in which it is difficult to know what one should do; that the ultimate beneficiary of ethical analysis is the person who, in one of these situations, seeks rational ground for the decision he must make; that ethics is therefore primarily concerned to find such grounds, often conceived of as moral rules and the principles from which they can be derived; and that meta-ethics consists in the analysis of the terms, claims, and arguments that come into play in moral disputation, deliberation, and justification in problematic contexts. (1986: 14)

Contemporary philosophical ethics is preoccupied with justifying correct answers to moral problems, and correct answers are those that can be derived from moral principles or rules. This method provides what Pincoffs calls the "rational ground" for a moral decision. We recognize it as the formal logic-based or analytic approach to ethics that we introduced at the outset.

"Quandary ethics," as Pincoffs dubs this logic-based, formal-reason brand of morality, is alien to much of moral philosophy: "Plato, Aristotle, the Epicureans, the Stoics, Augustine, Aquinas, Shaftesbury, Hume, and Hegel do not conceive of ethics as the quandarists do" (1986: 14). For Aristotle, for example, "In ethics the leading question concerns the best kind of individual life, and the qualities of character exhibited by the man who leads it.... Moral problems are given their due but are by no means stage-centre. The question is not so much how we should resolve perplexities as how we should live" (Pincoffs 1986: 15).

The women whom Lippman and Fraser studied certainly are trying to solve a problem, but they are not approaching it formally in the manner of a Pincoffs quandary. Rather, their age-old problem is precisely how they should live—they are deciding what they need to do to flourish. What is the best life for them, and for their child should they have one, and for their families? What qualities of character does that kind of life require? Are their families and they capable of exhibiting those qualities? To be sure, there is a particular problem they must resolve, but that problem is embedded in and subsidiary to their more sweeping moral inquiry. As these women astutely realized, their problem is satisfactorily resolvable only within the broad contexts of their inquiry, and their inquiry engages their entire being, not just a narrow calculational capacity. Quandary ethics, whether operating through deduction from principles or by application of the MEU principle, does not—cannot—appreciate the nature and momentousness of their task or all that it takes to grapple with their task. Re-reasoning ethics is about getting these matters right.

But what does it mean to "re-reason" ethics, and how can ethics be "re-reasoned"? Morality has to be rational. If morality is not rational (and not, we suppose, imposed by divine force or equivalent), then it becomes simply a matter of personal opinions or idiosyncratic preferences. Applying a priori moral principles or rules to the facts of problems to derive conclusions about what ought to be done can claim to be rational because it relies on the certain truth of the principles and utilizes the paradigmatic rationality of deductive logic and mathematics. Yet the deliberation of the women after genetic counseling, as we have seen, involves much more than this: it concerns no less than reimagining everyone and everything involved; it is complex, contextual, and riven with vagueness and uncertainty. Is their decision making rational? Not in the terms of the logic-based approach with its formal reasoning that renders deduction and mathematics rational. But what else can rationality be? Pincoffs recognizes the philosophical incredulity that greets any proposal to look elsewhere for rationality: "To abandon the search for rationally defensible rules and principles is to abandon moral philosophy" (1986: 14). Any other conception of rationality becomes (literally) unthinkable. Kurt Baier provides a crisp moral illustration of this philosophical dogma in his explanation of "why we can often tell immediately (as if by a moral sense or by intuition) whether a given course of action is right or wrong": "We do so by our reason, by subsuming a given course of action under a given principle" (1958: vii–viii).

The reduction of rationality to formal reasoning is so pervasive that even Lippman and Fraser echo it. They note that data from descriptive studies of "real-life" decision making indicate that when people make complex choices, for example, selecting a spouse or a career, they more often than not violate the normative model that is regarded as appropriate. Lippman and Fraser attribute this quirky "real-life" decision making not to the abandonment of rules but to the invocation of unauthorized rules: "The choices made are neither capricious nor random, but they follow rules other than maximizing expected utility" (1979d: 327–328). Their explanation assumes that decisions can be rational only if they are rule-governed; decisions that are not rule-governed are arbitrary or capricious. Because the deliberation of the women they studied goes well beyond rule-bound reasoning, the rationality of their decision making cannot be understood in the conventional way. Lippman and Fraser concede that: "the notion of rationality may not only be inappropriate, but may also limit our understanding of counselee's behavior and the influence of genetic counseling on this behaviour" (1979d: 325). And they go further: "Parents' perceptions of their situation as uncertain in a sense obliges them to follow what might be called 'a-rational' methods in resolving their choice problems. Rationality becomes irrelevant" (1979d: 330). That drastic, regrettable conclusion follows only if rationality is restricted to formal reason—to the exercise of deductively applying principles and rules.

But why accept the dichotomy, "either rule-bound and rational or not rule-bound and a-/irrational"? Many intelligent problem-solving activities are not rule-bound. Ironically, the construction of logic proofs itself, for example, is provably not a rule-bound process for nondecidable logics (most of them), yet it is a thoroughly skilled, intelligent activity. Creating a new logic still more so. Are mathematicians and philosophers who devise proofs or create logics acting a-/irrationally? Supposing that would divorce rationality from intelligence, thereby gutting both. Accepting that dichotomy commits what Richard Wasserstrom, writing about judicial decision making, insightfully calls "the irrationalist fallacy":

> The question of whether an argument is formally valid is confused with the question of whether there can be good reasons for believing a proposition to be true or false. It is recognized quite correctly that the canons of formal logic have a decidedly limited applicability. It is inferred quite incorrectly from this that all the questions which cannot be settled by appealing to formal logic cannot be settled in any manner which could be called "reasonable" or "logical." An appeal to formal logic is equated with an appeal to criteria of rationality or reasonableness, and it is concluded that because the courts cannot have used formal logic to select or formulate legal premises, the courts cannot have appealed to any rational or objective criteria when engaged in these undertakings.... It may not

make much sense to describe the judicial decision process as a completely deductive one. But it makes even less sense to insist that for this reason courts could not (and should not) employ a procedure or a set of procedures that permits of some kind of reasoned justification for the judicial decisions reached by those courts. (1961: 23–24)

Likewise, it makes no sense to regard the decision making of the women whom Lippman and Fraser studied as a-/irrational. Their situations are saturated with uncertainty and ambiguity. They must construct a process that allows them to manage that uncertainty and that ambiguity, to frame their problems and the information they receive in ways that are meaningful to them, and to arrive at a conclusion about what they should do. They accomplish all that, and the outcomes of the processes they devise display an immeasurably richer understanding of their situation and of themselves and their family in that situation than what they had at the outset and that any formal calculation could provide. At the same time, they dramatically improve the practical adequacy of their assessments. Moreover, their deliberation confers an enhanced capacity to tackle other such problems in the future. They have learned, and while doing so they have learned how to learn about resolving such deep problems. The processes these women use to structure their problems and to work their way through them are dynamic, creative, and contextual. And a bewildering array of diverse considerations are rooted sensitively and systematically in those processes, thereby belying charges of arbitrariness and capriciousness. Incorporating what these women have to teach into the practices of genetic counseling would improve both the intelligence and the moral quality of counseling processes.

At the very least the deliberation of these women is a model of powerfully intelligent, nonformal problem solving. In fact, to deny rationality to their problem-solving process would be to deny it to all the fundamental dimensions of our lives. How to throw a good party, settle a political disagreement, develop a new technology, solve a crime, design a unique piece of jewelry or a new kind of bridge, judge a legal case, become a saint, research a new scientific domain, or constructively parent a child? They are all problems that, like the problems of these women, are complex, multifaceted, multivalued, highly contextual, and mutually interactive, and they all begin ill-defined in form, method, and boundary. Let us call them deep problems. They are just as impenetrable to narrow inferential/calculational rationality as are the women's problems. And, as we shall see in chapters 3 and 4, scientists use this very kind of process when approaching deep research questions. Thus to call these women a- or irrational is tantamount to denying rationality to virtually the entire substance of intelligent life.

So what are we to say about the rationality of these women? Essentially there are only three possible responses. The women can be held to be: (A) intelligent, but a- or irrational because they do not use deduction to apply formal principles, for example, MEU; (B) a- or irrational, so unintelligent; or (C) intelligent, so rational. Option B precludes identifying intelligence with deep problem solving, simply to preserve the undefended assumption that rationality must be deduction from principles or rules. Ironically this would cut intelligence adrift from all but the most superficial problem solving. Option A sunders rationality from intelligence. We see no practical or theoretical benefit in doing that. To the contrary. The current sharp divisions among conceptions of intentionality, rationality, and intelligence— namely, intentionality modeled on linguistic reference, a late-evolved trait in human beings; rationality modeled on simple deduction from principles or rules; and intelligence modeled on formal, for example, computational, problem solving—offer an implausible biological picture of the evolution of isolated mental tricks. As opposed to this account, the model of deep problem solving represented by the women's methods, as we shall see, a model widely applicable in science, law, design, and elsewhere, supports a more plausible evolutionary model that understands intentionality and rationality as but two complementary facets of a single, developing problem-solving and decision-making capacity. Thus we conclude that option C thoroughly deserves our allegiance.[6]

For all these reasons, ethics needs to be re-reasoned. To that end, the regnant conception of rationality needs to be expanded beyond the narrow equation of rationality with formal reasoning to account for the rationality of the deliberation of the women who received genetic counseling, along with much of the deliberation that suffuses our lives. Developing that broader conception of rationality and articulating its wide-ranging consequences for ethics are the complementary projects of this book.

1.4   Four Lessons

Four key lessons emerge from the case presented in this chapter.

The *first* and most obvious lesson is that the dominant conception of rational procedure constrains roles and outcomes in problem-solving or decision-making situations. The genetic counselors' implicit understanding of and commitment to formal rational decision making clearly constrained both their counseling role (to just providing empirical outcomes and probability data) and the women's role (to just combining that empirical information with their personal evaluations of outcomes and then cost-benefit maximizing to reach a decision). Constraining behavior is inherent to any account of normativity, so constraint per se is not the issue. Rather, what this example highlights is that if we judge that the decision-making approach to which a normative account constrains us is inadequate, as these women did, the issue that properly arises is that of whether there is something wrong with that normative account (and not with the women).

In short, *second*, we can reasonably test the adequacy of a normative account by evaluating the particular decision-making approach it requires against our knowledge of the alternative problem-solving processes we are able to undertake.

And if there is an alternative approach that is judged superior, as happened with these women, then, *third*, it can be rational for the redesign to reciprocally include revising the nature and role of the rational procedure within which the superior approach is embedded. Then the issue that directly arises is what alternative normative account might better include within its obligatory constraints the decision-making approach considered most valuable. In their decision making these women identified and focused on much larger issues than what logical procedures address. They were concerned about what the design of a flourishing life meant for them in all their present circumstances.

So we conclude, *fourth*, that we will not be able to pass to the primary concern of designing a life proper without reconsidering the nature of rational ethical process, that is, of an appropriate problem-solving/decision-making approach.

Twin themes that structure the entire book emerge from these lessons. (1) We want to develop an alternative, more expansive conception of reason (alternative, that is, to the conception of reason as formal calculus) that is sufficiently rich to encompass all the procedures through which we humans can intelligently design our lives and our world to promote human flourishing in all its aspects. (This alternative account should describe a process that is itself learnable and improvable as we go about those very tasks.) (2) We aim to develop a sufficiently rich account of the proper scope and methods of rational ethics in this larger life-design task and show how these methods can lead to approaches and outcomes that are superior to those yielded by the methods of orthodox moral philosophy, indeed throughout the book to illuminate important cases whose resolutions are neglected because they are not representable within the logic-based approach.

2 The Difficulties of Applied Ethics

Bioethics, like its counterparts in the ethics of law, business, engineering, journalism, and other domains, generally has been regarded as applied ethics. In the first edition of their influential text, *Principles of Biomedical Ethics*, Tom Beauchamp and James Childress promulgated this view by characterizing bioethics as "the application of general ethical theories, principles, and rules to problems of therapeutic practice, health care delivery, and medical and biological research" (1979: vi–vii). Over six subsequent editions, Beauchamp and Childress have refined and embellished their method in response to a panoply of criticisms, while bioethics, despite all these changes, continues to be widely portrayed and practiced as a matter of applying "general ethical theories, principles, and rules" to concrete moral problems.

The persistence of the applied ethics model is easy enough to understand. In practice morality shades seamlessly into etiquette, law, or practical convenience, and the attempt to demarcate the distinctively moral becomes problematic. Moral norms might be differentiated from rules of etiquette in terms of their greater importance, and from legal rules in terms of their informal origins and the kinds of social sanctions imposed when they are violated, but these distinctions are anything but sharp and intrinsic.The only alternative to justify norms as genuinely and distinctively moral, it is widely accepted, is to organize them into a distinctive and consistent system of principles or rules to form a moral theory, perhaps through axiomatization. The only sure way to accomplish this theory construction— and the one typically adopted—is to appeal to a priori insight. Then, right actions can be logically deduced by applying the theory to the empirical features and circumstances of situations, thereby rendering practical morality a rational, objective enterprise. This conception of morality is not only pleasingly tidy, it also makes morality an autonomous discipline.[1]

The conception of bioethics as applied ethics incorporates this familiar philosophical account of what morality is and how moral justification proceeds. Bioethics is, in this view, a norm-constituted theory, and in what has become the canonical understanding of bioethics, that theory comprises Beauchamp and Childress's four fundamental principles: a principle of nonmaleficence, a principle of beneficence, a principle of autonomy, and a principle of justice.[2] Bioethical decision making is a matter of applying these principles and subordinate norms derived from them, such as a principle of informed consent, to moral quandaries in health care.

Despite the hegemony of this approach, conceiving morality in this way is, as we saw in chapter 1, far too simple. The methodological ferment in bioethics that followed the publication of the first edition of Beauchamp and Childress's *Principles of Biomedical Ethics* and that continues today testifies to the weaknesses and limitations of applied ethics thus understood. This chapter uses vignettes to depict core issues in bioethics that illustrate those failings in a practical, concrete manner. The vignettes display the profound limitations and deep problems that characterize any such model of applied ethics and the moral philosophy from which it is derived.

2.1 Indeterminacy: What Does a Norm Mean?

To apply a norm—a principle or a rule—one must know what it means. For norms to be norms, that is, to be directives applicable to an array of cases and not just commands for specific cases, they must be appropriately general. Moral principles and rules assign a normative status—for example, permissible, obligatory, or forbidden—to general classes or types of act. But then the relevant classes or types of act to which they are applicable must be specified; that is, criteria must be provided for identifying the particular acts that fall within the ambit of a class or type to which a norm applies. For instance, characterizing assault as including any plunging of a knife into a living human body will not do since it includes medical surgery to save lives and plastic surgery to save face; but it ought to include a surgeon's deliberate, unnecessary surgery performed in revenge.

The world is richly complex, dynamic, and protean; the general terms of a natural language are vague; and we are finite and fallible. We cannot understand everything about the world, and we cannot anticipate the multifarious ways in which our aims, goals, and values will interact and change and will impinge upon the world. So classes of acts never can be rigorously, precisely, and exhaustively defined, and criteria for identifying the acts that those classes comprise never can be correspondingly detailed. Problems of classification inevitably arise, and those problems create indeterminacy in the application of moral rules and principles. Consequently, judgments about when and how to apply these norms have to be made. The problem of classification plagues all of applied bioethics but is particularly compelling with respect to two of its most developed and contentious topics.

2.1.1 The Meaning of "Euthanasia"

Discussions of the moral and legal permissibility of euthanasia can quickly founder on uncertainty and disagreements about the meaning of the term. Animals are "euthanized" with a lethal injection. Must euthanasia for persons be similarly "active," or can it be "passive"? If there is a difference between active and passive euthanasia, is that because active euthanasia is killing, whereas passive euthanasia is allowing to die? Also, conceptual unclarity, and its attendant moral confusion, create practical problems of classification. Is the withdrawal of life-sustaining treatment, such as a mechanical ventilator, an instance of euthanasia? Is the withholding of life-sustaining treatment, such as antibiotics for pneumonia in a severely demented resident of a nursing home, an instance of euthanasia? The first vignette poses another classification problem, one that concerns the potential use rather than nonuse or discontinuation of an intervention.

Vignette 2.1   Ms. B

Ms. B, who is 30 years old and single, was told by her doctor on Christmas Eve that she had an ovarian mass that might be cancer. Supported by her sister, her only sibling, and her mother and father, she struggled with the uncertainty, anxiety, and fear of this possibility throughout the holidays. Afterward she was diagnosed as having a serious form of ovarian cancer that required immediate surgery and chemotherapy. Over the next two years Ms. B endured bouts of radiation therapy, more chemotherapy, and more surgery, intermingled with times when she could feel her life still was her own. At the end of those two years, Ms. B had a colostomy. She was expected to live for only a few more months.

Soon thereafter, Ms. B became bedridden at home and clearly was dying. She asked Dr. M, her palliative care physician, whether he remembered all the talks they had had over the past few months about how afraid she was to die struggling for breath. Ms. B reiterated that her physical pain was well controlled, but she stressed that her breathing was getting harder and harder, which scared her and made her breathing difficulties even worse. Dr. M assured her that he did remember, that he would not abandon her, and that he would not forget her wishes. Dr. M promised Ms. B that he would sedate her if her breathing difficulties could not be controlled. Ms. B assured Dr. M that the morphine helped her a lot, but she said she knew she was taking more and more to control her symptoms, and she felt she would need to be sedated soon. She added that she knew she was dying and she was not afraid to die. She knew she was going to a better place, and she just wanted to sleep and be at peace before getting there. She no longer wanted any food, and she was taking only sips of water and occasionally a small piece of popsicle.

One day a good family friend visited Ms. B, and they discussed her fear of suffocating and what would be done if that were to occur. The friend commented that he understood the plan and that he was a strong supporter of euthanasia. Ms. B was startled and nonplussed, and she became anxious and upset. After her friend left and Ms. B told her family what he had said, they became distraught and called Dr. M. The family explained that they were very religious and that although they appreciated everything medicine could offer, they did not want Dr. M to "euthanize" Ms. B if her breathing became difficult. Dr. M was shocked by the suggestion.[3]

Terminal sedation is used in palliative care[4] to relieve intolerable pain or suffering that cannot be controlled. Ms. B's pain is well controlled. Ms. B is not afraid of death. What Ms. B is afraid of is how she might die. She fears that when her breathing difficulties worsen, she will start gasping for air and feel that she is suffocating. That fear causes her anguish and her suffering. To avoid that suffering, Ms. B wants to be sedated to unconsciousness. She is eating only small bites of a popsicle and drinking only small sips of water, so once she is unconscious, all life-sustaining interventions will be withheld, including artificial nutrition and hydration. She will die of dehydration, starvation, or a complication or deterioration of her underlying terminal condition.

Ms. B, her family, and Dr. M agree about the governing norms: relief of suffering is morally obligatory, euthanasia is morally prohibited. This would seem to make the situation a straightforward case of applying these rules in the standard applied bioethics model. But there remains the difficulty of how to apply the euthanasia norm to their situation. Is terminal sedation an instance of euthanasia? Different answers to that question create moral turmoil. The family friend, who strongly endorses euthanasia, thinks it clearly is. Dr. M, whose shock shows that he strongly rejects euthanasia, thinks it clearly is not. Ms. B and her family, who also strongly reject euthanasia, are upset that it might be. If that matter of classification were settled, however, a solution to their moral quandary, according to the applied ethics model of bioethics, would follow swiftly and surely. A statement about whether terminal sedation is or is not euthanasia would be subsumed under the governing norm, and a conclusion about whether terminal sedation is morally permissible for Ms. B would be logically deduced. But how is that crucial matter to be settled? Is what appears to be a separate, well-defined problem of semantic classification—what does "euthanasia" mean?—really so delimited and independent of moral considerations?

Notice, for instance, that the question already has changed. The question no longer is about only terminal sedation, but terminal sedation in combination with the withholding of all interventions that might prolong Ms. B's life. Does this more complex plan make her envisaged death a form of euthanasia? The term "euthanasia" comes from Greek, where, taken literally, it means a good death (*eu* = good + *thanatos* = death). In that sense the mode of death proposed for Ms. B is, if not a good death, at least a better death than a suffocating death. Over two millennia, however, and more recently with all the technological advances in medicine that enable life to be extended well beyond previous limits, "euthanasia" has acquired complicated moral, legal, and social meanings. The meaning of "euthanasia" has become implicated in and shaped by the moral controversies it creates. Two prominent moral positions that address issues about when and how life may be ended have particularly influenced the understanding of euthanasia. One is the distinction between active and passive measures, and the other is the doctrine of double effect.

The difference between an active mode of hastening or causing death and a passive mode of hastening or causing death is, for many people, morally crucial. Active measures, such as administering a lethal injection at the request of a patient in the case of voluntary active euthanasia (VAE), are deemed morally impermissible, whereas passive measures, such as withdrawing or withholding life-sustaining therapies, are deemed morally permissible. The putative moral difference turns on what appears to be substantially different methods of causation. Whereas in VAE a doctor directly causes the death of a patient by administering an injection, in withholding or withdrawing life-prolonging interventions it is the patient's underlying disease that causes death. In the former, human intervention causes death; in the latter, nature causes death. That distinction can be difficult to sustain practically, however, because doctors take a series of active steps in withdrawing a patient from a mechanical ventilator. The distinction also is difficult to sustain theoretically. Patients who are terminally ill can end their lives by deciding to voluntarily stop eating and drinking (VSED). As Timothy Quill, Bernard Lo, and Dan Brock describe it, "With VSED, a patient who is otherwise physically capable of taking nourishment makes an active decision to discontinue all oral intake and then is gradually 'allowed to die,' primarily of dehydration or some intervening complication" (1997: 2099). VSED can be regarded as a passive measure, along the lines of withholding or withdrawing life-sustaining interventions, in which the cause of the patient's death is an underlying disease. But as Quill, Lo, and Brock object, "the notion that VSED is passively 'letting nature take its course' is unpersuasive, because patients with no underlying disease would also die if they stopped eating and drinking. Death is more a result of the patient's will and resolve than an inevitable consequence of his disease" (2102).

Might the active-passive distinction nevertheless allay the worries that Ms. B and her family have about terminal sedation? Terminal sedation could be regarded as morally permissible because "the administration of sedation does not directly cause the patient's death and because the withholding of artificial feedings and fluids is commonly considered passively allowing the patient to die" (Quill, Lo, and Brock 1997: 2102). Doctors and nurses, however, probably would consider the ongoing administration of the drugs needed to sedate a patient to unconsciousness to be an active process; moreover, the suggestion that, in terminal sedation, nature is taking its course is questionable because the patient often dies of dehydration from the withholding of fluids, not of the underlying disease (2102).

What about the doctrine of double effect? The doctrine of double effect is a technique for morally evaluating courses of action that have two consequences, one that is desirable and another that is undesirable. The suggestion that terminal sedation is a kind of euthanasia might have prompted Ms. B and her family to appreciate that aggressive sedation could depress Ms. B's respiration and thereby cause her to die sooner than she otherwise would without it. In that event, terminal sedation would have a double effect: the desirable effect of preventing suffering and the undesirable effect of hastening death. Would the latter make it morally wrong?

A central component of the doctrine of double effect is the intention with which one acts. For a course of action to be morally permissible, the desirable outcome must be intended, and the undesirable outcome must be unintended, even though it is foreseen, perhaps with a high degree of probability. It could be argued that the intent of administering terminal sedation is relief of suffering and that the acceleration of death, although foreseeable and highly probable, is unintended. But terminal sedation is accompanied by the withholding of life-sustaining measures, including artificial nutrition and hydration. With both, the dying process usually takes days, if not weeks; without withholding life-extending measures, the process would be substantially prolonged. When withholding artificial nutrition and hydration is added to the administration of terminal sedation, however, it is difficult to claim that only relief of suffering is intended, as Quill, Lo, and Brock explain:

> Although the overarching intention of the sedation is to relieve the patient's suffering, the additional step of withholding fluids and nutrition is not needed to relieve pain, but is typically taken to hasten the patient's wished-for death. In contrast, when patients are similarly sedated to treat conditions like status epilep-

ticus, therapies such as fluids and mechanical ventilation are continued with the goal of prolonging life. (1997: 2101)

Were withholding fluids and nutrition to be added to Ms. B's plan for terminal sedation, as a reasonable extension of her current behavior and as a reasonable thing to do in any event, the intent of that combination would be not just to end her suffering but also to hasten her death as a means to that end. So the friend of the family would be right.

Although the moral distress of Ms. B and her family is caused by the suggestion that terminal sedation is a form of euthanasia, asking whether terminal sedation is an instance of euthanasia as a preliminary to applying a moral norm does not help them grasp their moral situation more clearly or address their fears and worries. Like many general terms, especially those described as "essentially contested" (Gallie 1956), "euthanasia" does not have a rigorous, precise, unequivocal meaning. Deciding whether terminal sedation is a form of euthanasia cannot be a straightforward, determinate matter of applying a concept, as happens, for example, when a class of elementary school students is given a sheet of geometrical figures and told to color the triangles. The meaning of "euthanasia" is dynamic, shifting over time and across circumstances, and it is influenced by moral, legal, and social controversies, most recently, by ones about withholding and withdrawing life-sustaining treatment and legalizing physician-assisted suicide and voluntary active euthanasia.

There is another problem. The moral perplexity and distress created by Ms. B's situation focuses on the notion of euthanasia because it is assumed that the relevant governing norm concerns the moral status of euthanasia. That assumption can be questioned. Quill, Lo, and Brock, for example, reject an approach to determining the moral status of voluntarily stopping eating and drinking, terminal sedation, physician-assisted suicide, and voluntary active euthanasia that is predicated on applying the active-passive distinction or the doctrine of double effect. For them, what is morally important, with respect to competent patients, is the ethical principle of respect for autonomy. None of the four ways of ending life would be morally permissible, they emphasize, without the voluntary and informed consent of the patient. Ms. B is competent. Ms. B has discussed her fears with Dr. M, and they have come to an agreement about how to treat her impending breathing difficulties. She has made a voluntary, informed, considered decision about terminal sedation, and she has sought Dr. M's reassurance that he will keep his word. Respect for the autonomy of Ms. B requires that Dr. M provide terminal sedation.

This moral alternative rejects the norm that initially is assumed to govern and structure Ms. B's plight and reverses the applied judgment. So what is the proper moral approach here? Is a decision about terminal sedation for Ms. B a matter of the moral permissibility of euthanasia or a matter of respect for autonomy? On the basis of their analysis, Quill, Lo, and Brock conclude that "TS [terminal sedation] and VSED [voluntary stopping eating and drinking] are more complex and less easily distinguished ethically from PAS [physician-assisted suicide] and VAE [voluntary active euthanasia] than proponents [of VSED and TS] seem to realize" (1997: 2101). Trying to identify relevant similarities and dissimilarities among those four ways of ending life and then basing moral assessments on the results of those comparisons is, in their view, misguided. For them the common and central moral issue posed by all four, with respect to competent patients, is whether the patient's decision is voluntary and informed. Their approach takes us back to where we began. Like "euthanasia," "autonomy" is a general term. What does autonomy mean, and how are particular decisions of particular patients about particular problems to be classified as autonomous or nonautonomous?

2.1.2 The Meaning of "Autonomy"
In the second vignette respect for autonomy becomes the focus of life-and-death decision making.

Vignette 2.2   Ms. F
Ms. F, who is 28 years old and has cystic fibrosis, had a lung transplant two years ago. She began experiencing problems with her new lung several months after she was discharged from the hospital, and she subsequently had a heart-lung transplant, after which she was able to return home. A month later she had to be readmitted to the hospital because the second heart-lung transplant also was unsuccessful. The lung transplant team rejected the possibility of a third lung transplant because it was considered to be medically futile. Ms. F was encouraged to consider a hospice program in which, she was assured, her medical problems would be treated and palliative care would be provided when necessary.

Even though transplantation of a third lung in the same patient is rare, the conclusion that another lung transplant for Ms. F would be futile was not unanimous. A physician on the transplant team, who was about to join the staff at another hospital to direct their new lung transplant center, thought that Ms. F might be able to survive another transplant, and he communicated his opinion to her. If she wanted to be considered for a third lung transplant she could be transferred to the other hospital. Ms. F faced an agonizing decision. She knew that if she remained where she was, death would be imminent. She also knew that if she wanted another chance, she would have to leave the hospital where the lung transplant team had become, over the previous two years, a surrogate family to her.[5]

The centerpiece of contemporary bioethics is a principle of autonomy that authorizes patients to make decisions about their health care in terms of their own beliefs and values. Cognizant of this principle, the hospital staff request an ethics consultation for Ms. F. They are willing to respect whatever decision Ms. F makes, provided they can be confident that her decision is genuinely autonomous.

How is a clinical ethicist to make that determination? What does it mean for a decision to be autonomous? Bruce Miller has neatly distinguished four components of autonomy: autonomy as *free action, authenticity, effective deliberation*, and *moral reflection* (1981). An action is *free* when it proceeds from a decision that is intentional and voluntary, that is, not the result of coercion, duress, or undue influence; *authentic* when it is consistent with a person's attitudes, values, and goals, that is, "in character" for that person; the result of *effective deliberation* when the person deciding what to do is aware of the courses of action that are available and the likely consequences of these alternatives, evaluates them, and makes a choice based on that evaluation; the product of *moral reflection* when a person scrutinizes the values that have been inculcated in him or her, accepts a set of values as his or her own based on that critical examination, and then proceeds to make decisions based on those values.

Given Miller's analysis, the clinical ethicist first must determine whether Ms. F's decision is autonomous in each of the four senses. *Free action*: Whatever Ms. F decides would be intentional, but would it be voluntary? Ms. F is torn not just between her options, but between her two families. Her parents and her sister support the decision of the lung transplant team, but her boyfriend and his family want her to try again. There is pressure in both directions, so no matter what Ms. F decides, she will be doing something that somebody close and important to her believes is wrong. Do the competing pressures somehow cancel out, or should they simply be accepted as a fact of life? If the latter, what distinguishes troubling yet morally benign pressures from pressures that render decisions unfree—pressures that constitute undue influence or duress? In this regard, consider how important the lung transplant team has become to Ms. F. Members of the team have been caring for her for months at a time over a period of two years. Ms. F understands how dependent she is on them, and she recognizes how she resists the authority they have and the power they wield over her. One of her journal entries reads: "I woke up in a horrible mood. Sick of being here in the same place, no patience for anything, being a whiner. All it does is make things harder. Jim is my nurse. I know I must piss him off. I got a big dose of Morphine and that just about knocked me out— so now I'm laying dizzy and paying for it. I bring this on myself I think" (Marshall 2001: 142). Ms. F's emotions are roiling, and her moods can be dark. She recognizes her intemperate reactions, and she knows her bad moods have consequences: "a big dose of Morphine" in this case. Given her dependencies and vulnerabilities, there is strong pressure on Ms. F to be compliant, to be a "good" patient. Is that kind of pressure morally innocuous, or does it raise concerns about the autonomy of her behavior and her decisions?

Effective deliberation: Ms. F agonizes over her decision, strives to understand her predicament, and seeks guidance in her religious convictions. Does that qualify as deliberation, and if so, is it effective deliberation? The model of effective deliberation in bioethics is exemplified by the informed consent procedure. To obtain valid consent for treatment from a patient, a health care professional must inform a patient about the nature of the problem, alternative forms of treatment, the likely benefits and risks of those alternatives, and the likely outcome of doing nothing. The patient then is supposed to assess the consequences of these options in terms of his or her own values and make a decision on that basis. To facilitate this kind of deliberation, the clinical ethicist could construct a decision tree and guide Ms. F through a detailed consequentialist process of reasoning. Chapter 1 already has provided a telling lesson about the moral traps in this approach. In that case effective deliberation began by rejecting the whole apparatus of benefit-cost maximization. And here it turns out that such an exercise might not be meaningful to Ms. F because she too does not structure her problem in that way.

Ms. F's thinking is more metaphorical. During the consultation, the clinical ethicist reads a poem Ms. F has written in which she compares her body to a bicycle with stripped gears. A portion of the poem goes: "... but your bike is starting to fall apart, the gears stripped, leaving you with the exact one you don't need. ... They fix the gears and tell you go on ... keep going it's gonna get easier, something's wrong though cause you fall again, hard and this time the gears can't be fixed. ... A lot of people say never give up. Others tell you to do as ... best you can. How do you learn to accept? Or do you fight 'till the absolute end'?" (Marshall 2001: 144). Bikes break and are taken to mechanics to be fixed. Bodies break and are taken to surgeons to be fixed. Indeed, there is a surgeon ready and willing to try to repair Ms. F's broken body. What should the clinical ethicist make of Ms. F's metaphor? In particular, should the clinical ethicist challenge the mechanistic, reductionistic image of persons and their bodies that it conveys and point out the kind of bias this assumption introduces into Ms. F's decision making? Ms. F is thinking about her situation and trying to work her way through her problem, but is she engaging in the kind of deliberation required for her decision to be autonomous? What should the clinical ethicist conclude if her process of deliberation does not match the consequentialist reasoning presupposed by the exercise of obtaining informed consent?[6]

Authenticity and *moral reflection*: Assessment in these two categories is even more difficult because it requires historical and personal information about the social, psychological, emotional, and spiritual dimensions of Ms. F's life. Would a clinical ethicist be able to establish the trust and intimacy required just to broach such topics? Would Ms. F be willing to allow her family and friends to talk about private aspects of her life with a clinical ethicist? Would a clinical ethicist have the training and the skills needed for that kind of interviewing? Although the senses of free action and effective deliberation convey the kind of control autonomy protects, the senses of authenticity and moral reflection convey the deeper meaning of autonomy that gives that kind of control its moral importance. Satisfaction of the former two senses establishes the existence of control, but the latter two senses are necessary to understand whether and how a person wishes to exercise control.

Were the clinical ethicist able to sort out these indeterminacies, another problem could arise. Ms. F's decision might be autonomous in some senses but not autonomous in other senses. Are all four senses required for a decision to be autonomous? Do mere numbers count, or are some senses more important than others? How, in other words, is the clinical ethicist to reach a conclusion about whether Ms. F's decision is, all things considered, genuinely autonomous?

Miller recognizes that "what sense of autonomy is required to respect a particular refusal of treatment is a complex question" (1981: 27). Two considerations that Miller cites as relevant to answering that question are the nature of the relationship between the person making the decision and the person examining the autonomy of the decision, and the seriousness of the decision. What kind of relationship could Ms. F have with a clinical ethicist whom she did not ask to see, of whose existence she probably had no knowledge, and whose line of work undoubtedly would be baffling to her? And even if a close relationship between Ms. F and the clinical ethicist were to emerge quickly, how would that be relevant to assessing the autonomy of Ms. F's eventual decision? Perhaps the relevant relationship is not between Ms. F and the clinical ethicist, though, but between Ms. F and the lung transplant team. The clinical ethicist is merely an agent of members of the team with whom Ms. F has established close relationships. The relevance of close relationships to determining the autonomy of decisions remains unclear, however. Are there different tests for relationships with strangers and relationships with intimates, with lay persons and professionals? If so, why and how do those tests differ? Is a bioethicist determining autonomy like a consulting psychiatrist determining competency, or is autonomy even more elusive? Relationships are morally important, but how they are relevant to applying the principle of respect for autonomy is not clear.

Miller's other concern about the seriousness of the decision is straightforwardly satisfied, though. Ms. F is making one of the most important decisions she ever will have to make. But what does that mean for an assessment of the autonomy of her decision? Does it mean that the criteria for the four senses of autonomy should be applied more rigorously and stringently? Is Ms. F's decision making being scrutinized because the lung transplant team is worried that she might make a mistake? Indeed, in Ms. F's circumstances does it make sense to talk about a "correct" decision?

Miller's two considerations are important because they recognize the moral relevance of the particularity of persons and relationships and the details of situations and settings. But contextualization raises serious problems for applied bioethics. When interpretations of the various senses of autonomy and assignments of priorities to those multiple senses become context-dependent, the claim that moral decision making is a matter of applying rules is hard to sustain because there would have to be second-order rules for interpreting the first-order rules about autonomy, in particular for determining which features of persons, relationships, situations, and settings are morally relevant. There also would have to be second-order rules for resolving conflicts between competing interpretations of the notion of autonomy. But then those second-order rules could need interpretation, and conflicts between those interpretations would have to be resolved, so there would have to be third-order rules to govern those decisions. Not only would an infinite regress of rules loom, but our moral insight has somehow to extend to all these new and increasingly abstract rules. Does it really? We suggest not, as indicated by the nearly universal absence of plausible such rules. Conversely, were such rules somehow conjured, they would become so numerous and so complicated that applied bioethics, even if it were still theoretically tractable, would become practically unmanageable.

Moreover, like the first vignette, this one raises the issue of how the governing norm is determined. Is Ms. F's problem one of questionable autonomy? In an early meeting with the clinical ethicist, Ms. F asked, probably rhetorically, "What is my choice? To live like this for the rest of my life [while gesturing to the monitors and tubes in her hospital room]? Or to try it again and see what happens? Is there a choice?" (Marshall 2001: 146). If Ms. F has no real choice, why the moral preoccupation with respecting her autonomy? Ms. F is trapped, and one has to have enormous sympathy for her, for how she agonizes, and for whatever she decides. Given her plight, how can a clinical ethicist wielding the limited resources of applied bioethics determine whether her decision is genuinely autonomous? That the ethics consultation focuses on the autonomy of her decision making is an artifact of the prevailing applied-ethics model of bioethics, not a perceptive, sensitive response to her dilemma. The job of the clinical ethicist is to help Ms. F, not just the lung transplant team. Recalling the genetic counseling case in chapter 1, the clinical ethicist might be able to do that effectively, not by focusing on value maximization, but by listening to and comforting her, by sensitively assisting her to explore her own perceptions and feelings about her situation, and by supporting her. Yet applied bioethics offers no such alternative. It is no surprise, therefore, that the clinical ethicist can do so little to help Ms. F make her agonizing decision about whether to try a third transplant.[7]

2.2   Conflicts: Which Norm Prevails?

When a moral theory contains more than one principle, conflicts between those principles inevitably arise, as, for instance, the conflict between relieving suffering and avoiding euthanasia for Ms. B in vignette 2.1. In contemporary bioethics the familiar issue of paternalism results from a conflict between a principle of respect for autonomy and a principle of nonmaleficence. After careful deliberation a competent patient might, for example, refuse life-sustaining treatment that a physician believes is likely to provide the patient with a reasonable quality of life. Should the physician respect the patient's autonomous decision or try to prevent what he or she regards as avoidable harm to the patient? The latter alternative is disparaged as being paternalistic because the physician would be assuming that he or she knows, better than the patient does, what is in the patient's best interest. Respect for autonomy has, at least in theory, become firmly ensconced in contemporary bioethics, even if that respect eventuates in harm that could have been averted.

Conflicts about the use of life-sustaining treatment can be even more agonizing, however, when the life at stake is that of an infant, as in the following vignette.

Vignette 2.3 K'aila

In the weeks after his birth at home, K'aila nursed well, grew, and gained weight. His skin and the whites of his eyes remained a little yellow following an episode of newborn jaundice, but K'aila's doctor was not concerned. When K'aila was still yellow almost three months later, however, he was examined in a hospital by a pediatrician. Preliminary tests revealed abnormal liver function, but additional tests did not identify the cause of the liver dysfunction. The diagnosis subsequently was narrowed to either biliary atresia or giant-cell hepatitis. There is no treatment or cure for either. K'aila's mother and father were told that the only way to prolong K'aila's life, as his condition deteriorated, would be to perform a liver transplant. K'aila's pediatrician explained that transplantation is the standard, accepted treatment for end-stage liver disease in North America. He cited survival rates that approached 80% to 85% with improved techniques and the use of the drug cyclosporine to prevent organ rejection.

K'aila's mother and father are Native Americans who from the outset had deep misgivings about a transplant. They struggled with their decision and prayed and meditated for months. They also acquired more information about transplantation. They learned that K'aila would have to take immunosuppressants for the rest of his life. They discovered that the estimated 80% to 85% survival rate was high compared with other sources that cited a 70% to 75% survival rate. They realized that these survival rates are for a period of one year and that survivors die at a rate of 1% to 2% each year thereafter, so that the survival rate at five years after a transplant is closer to 60% to 65%. They also discovered that immunosuppressive therapy renders the body vulnerable to infections, the most dangerous of which are viral. Chickenpox, for example, can be fatal to a child on immunosuppressants. And they learned about the side effects of the immunosuppressant drugs, which include progressive damage to the kidneys, cholesterol buildup, hormonal changes, neurologic disturbances, increased hairiness, and a heightened risk of cancer.[8]

K'aila's parents decided to refuse a transplant. K'aila's pediatrician was uncomfortable with their decision, and because they sensed he would have difficulty providing only supportive care to their son, K'aila's parents transferred his care to a different pediatrician. The original pediatrician reported K'aila's parents to the provincial Department of Social Services, which referred them to the Department of Social Services in a neighboring province. Officials of that province applied to the courts for custody of K'aila so that an assessment could be performed and K'aila could be put on the list for a liver transplant without the consent of his parents. The court upheld the right of K'aila's parents to make a decision that was within the bounds of conventional medicine and that they believed was in the best interest of their son.

At the most abstract level the disagreement between K'aila's parents and his pediatrician can be couched in terms of what is in K'aila's best interest. "Best interest" is, however, an amorphous notion. To K'aila's pediatrician it means that K'aila should be given a chance to live, even if his life would constantly be at risk and the quality of his life inevitably would be compromised by unending medical problems, complications, and side effects. To K'aila's parents it means that the suffering K'aila would have to endure is not worth the risky grasp on staying alive that it might allow.

Applied bioethics depicts this dilemma as a conflict between principles, one venerating the quality, the other venerating the quantity, of life. Which principle is more important in this situation? How can that be determined? For applied bioethics there would have to be a higher-level norm that governs that choice. But there are many such conflicts among principles, so there would have to be many higher-level norms, and we must expect that they also would conflict, requiring still higher-resolving norms, and so on.[9] How, then, does applied ethics determine which principle is right when there is a conflict? Intuitions might be invoked, but, as the vignettes illustrate, there are few such intuitions, and they are weak and/or fuzzy. More common is a resort to "balancing" or "weighing" principles, an approach that, as the Statue of Liberty symbolizes beautifully, turns out to be no more than a metaphor, lacking any substantive correlate of gravity to do the work. Moreover, all such approaches conceal the contextual judgments that are needed to assess how much importance norms have in the unique particular situations to which they are brought.

By portraying moral problems as conflicts between principles, applied bioethics fosters the impression that, in the likely absence of a dispositive higher norm, these conflicts either are stalemates or are settled only by appealing to subjective, nonrational considerations external to norms. Moreover, imposing this norm-constituted, norm-governed reasoning on moral problems simplifies and distorts the complex nature of real moral deliberation. K'aila's mother provides an illuminating and sensitive account of how she struggled to come to a decision:

> The more I learned, the more integrated my decision became. I kept coming back to the same choice, at several levels, until it became clear to me that I could not live with any other. Had I given in to pressure to accept intellectually an ill-defined statistical chance that K'aila might survive an unknown number of years, with an

unpredictable number of problems and side effects, I would have had to deny what I knew at other levels. I would have been torn for the rest of my life, knowing that I had made a decision that didn't feel right inside. (Paulette 1993: 15)

The reflection that was meaningful to K'aila's mother had nothing to do with numbers and rules. Her deliberation operated at several levels, so she was not thinking just about K'aila but also about how she was thinking. Moral deliberation requires that multilevel complexity and reflexivity: thinking about a problem and at the same time reflecting on that thinking. With respect to the former, for example, how open to be to others who are affected and how spiritual and medical considerations are to be related; with respect to the latter, how one is oriented toward the problem and how sensitive to be to the particularities of the situation. K'aila's mother reports that "it was in quiet moments with K'aila that I felt most lucid" (Paulette 1993: 16). One of those lucid moments occurred in a hospital ward when an intravenous line in K'aila's foot was making him uncomfortable:

At last I took him out of his crib and brought him to bed with me on a narrow cot. As he snuggled into my arms and began to nurse, he relaxed and fell asleep. I watched him sleeping peacefully, and it became clear to me what I had to do for him. I imagined him like a butterfly, so beautiful, indeed exquisite, and yet so fragile. With my palm open wide, he had alighted on my hand and now rested there a few moments, his wings shimmering in the light. I longed for him to stay forever, but I knew that at any time he might flutter his wings and be gone. The only way I could hang onto him would be to close my hand around him, to entrap him in my grasp. However, I knew that in doing so I would run the real risk of harming him; I might open my fingers again to find his wings broken and crippled, his colors faded and smudged. I knew in that moment that I must keep my palm open and let him take flight when the time came. (Paulette 1993: 16)

The reasoning that engages K'aila's mother—all of her, not just her intellect— is metaphorical. It is that reasoning that moves her and on the basis of which she decides what to do.

In this respect as well, the norm-governed reasoning of applied bioethics ignores the profound contextuality of moral deliberation. Moral delibera- tion is grounded in background epistemological and metaphysical assump- tions that both structure and give meaning to the deliberation. When confronted by the prospect of a transplant for K'aila, his mother and father "had a general and overwhelming sense that it would be an unnatural thing to do" (Paulette 1993: 14). K'aila's mother explains the metaphysical context of their belief:

Deep within me there is knowledge of the intimate relationship between the spirit and the body. One lives on long after the other has fallen away but, in this brief interval, the two are intimately connected. This knowledge provides the basis for many specific practices and observances in my daily life. One, for example, is the care with which I dispose of my hair when it has been cut. My hair is part of me, associated with my spirit, so I can't just allow it to be swept into the garbage. So it was that I found it deeply disturbing to consider taking out part of K'aila's body, given to him by the Creator, and discarding it in favor of part of someone else's body, associated with a different spirit. Such a radical intervention in his life, which not only would defy the physical laws of nature but also might disrupt his spiritual path, seemed utterly perverse. (Paulette 1993: 14–15)

Associated with these metaphysical views about bodies, spirits, and persons are epistemological views about what can be known and how it can be known. K'aila's mother articulates those as well:

I tried to explain these feelings to the pediatrician, but I'm not sure that he ever understood. These were not merely "religious taboos" or prohibitions that prevented me from making a "rational" choice. These feelings were manifestations of knowledge that was available to my spirit, long before it was ever available to my mind. The human spirit can reach horizons far beyond what the eyes can see and can perceive the path that lies ahead long before the mind can make sense of the signs that mark the trail. (Paulette 1993: 14)

That explanation undoubtedly was hard for the pediatrician to comprehend because K'aila's mother invokes an epistemology remote from and alien to the epistemology of medical training, scientific research, and formalist Western rationalism in general. The views and the approach of K'aila's parents probably do seem irrational to the pediatrician, and may seem irrational to many readers of this book, because they proceed from the mysteries of faith and feeling rather than the empirical measurabilities of science and technology. And "true believers" can indeed proceed dogmatically and uncritically, but it is not always unreasonable to involve matters of spirituality in decision making. What is reasonable and what is unreasonable will be multi-factored, context-dependent, and revisable.

Whether the decision making of K'aila's parents was rational was the key issue in the judge's disposition of the application to have K'aila committed to the care of the Minister of Social Services so that the Minister could authorize and consent to the transplant operation.[10] The judge recognized the nature of the choice that had to be made:

There is a 100% certainty of death for this child without the surgery. With the surgery there is approximately a 70% to 75% chance of survival in the first year, assuming the child receives a positive assessment and a donor organ. There is between a 60% to 65% chance of five-year survival. Survival necessarily entails an uncertain quality of life. The child will be very susceptible to infection. The child will suffer the surgery and the known side effects of the follow-up drug therapy. Death may be avoided and life may be prolonged in the short term. There are no guaranties. (140–141; page numbers in this section refer to the case cited in note 10)

The judge also recognized that medical opinion was divided. On the one hand, liver transplantation is a standard and accepted treatment for end-stage liver disease in both adults and children in North America. On the other hand, because of the success rates, the risks of the procedure and of the subsequent drug therapy needed to suppress the immune system, and the impaired quality of life, allowing a patient with end-stage liver disease to die with supportive care and treatment also is a legitimate, medically recognized alternative. The judge then made two important determinations.

The first concerned the scope of the issue. The judge recognized that the argument for a transplant is narrowly medical, and he appreciated that a decision about a transplant "cannot be reduced to mere mathematical probabilities" (143). The resistance of K'aila's parents, the judge also recognized, is grounded on broader practical, emotional, social, and psychological considerations for both K'aila and his parents. Those considerations, in the judge's view, were decidedly relevant: "It is fair to say that the parents have given a much more profound consideration to the other components in the decision. These other factors must necessarily be given considerable weight" (142).

The second determination concerned procedural matters and the nature and quality of the decision making of K'aila's parents. The judge found K'aila's mother to be "rational, objective, and balanced in her assessment of the situation" (140). He noted that she "has become thoroughly conversant with the available information outlining the benefits and risks in making a decision" (139), and he pointed out that she had met with two families who have children with transplants and had discussed the situation extensively with them and with various doctors, as well as the director of the transplant program to which K'aila would be referred. The judge concluded that K'aila's parents "are both intelligent individuals" (139) who after thoughtful and careful consideration have made a fully informed, considered decision to withhold consent for the transplant operation. Given all that, the judge found that their decision making was not unreasonable:

> The decision reached by the parents cannot be characterized as being overwhelmingly poor, inherently inept, ill-considered, lacking judgment, or simply unreasonable and unacceptable to society. Their decision is not of the quality that the state should be compelled to intervene. ... The parents' choice of care is not unreasonable in these circumstances. (142)

Strikingly, the judge quoted Hippocrates: "Hippocrates said, 'Life is short, the art is long, timing is exact, experience treacherous, judgment difficult.' This is just as true today as it was in 400 B.C." (142). The judge understood that a decision about K'aila's future required a difficult judgment. Given that understanding, the judge converted the issue from one of substance to one of process: how did K'aila's parents make their decision, and was their approach, in the circumstances, reasonable? The judge's inquiry was not purely procedural, however, because he recognized that there are substantive considerations that impose constraints on the decision making of K'aila's parents. The judge asked, "Can it be said that this is a clear case of rejection by the parents of the values society expects of thoughtful, caring parents for a terminally ill child?" and he answered in the negative. That question raises issues of substance and procedure and itself calls for judgment: for example, which values should be involved here, and when is a case of rejecting society's values a "clear" case? For present purposes, though, the important point is that the approach it adopts and the constraints it imposes are markedly not the approach and not the constraints of applied ethics. The next section provides another example of how the failings of applied bioethics prompt a shift from substance to process.

2.3   From Substance to Process

The manner in which the judge decided K'aila's fate is distinctive of, but not peculiar to, judicial decision making. Indeterminacies in, and conflicts between, norms—in law, in bioethics, and everywhere else—often make it impossible to straightforwardly derive a decision. Yet a decision has to be made. How does that happen? As in K'aila's case, the focus shifts from the right decision to making a decision in a right way. John Arras's analysis of the following vignette illustrates that shift in bioethics. When the norms of bioethics run out, as they quickly and inevitably do, the question of *what* to do is replaced by the question of *how* to decide.

Vignette 2.4   Mrs. Smith

Mrs. Smith, an 85-year-old resident of a nursing home, was transferred to the hospital for treatment of pneumonia. Although she has responded well to antibiotic therapy, her overall condition and prognosis remain grim. For the past three years her mental state has been steadily deteriorating due to a series of strokes, which have finally rendered her severely demented. She is now nonambulatory, incapable of sitting up in bed, and uncommunicative most of the time. When she does talk, her speech is completely incoherent and repetitive. Mrs. Smith shows no signs of recognizing or remembering her family and primary caregivers. The nurses in charge of her care assert that she appears to experience pleasure only when her hair is combed or her back rubbed.

During her recovery from the pneumonia, Mrs. Smith began to have problems with swallowing food. Following a precipitous decline in her caloric intake, her son and daughter (the only involved family members) consented to the placement of a nasogastric tube. Mrs. Smith continually pulled out the tube, however, and continues to resist efforts to reinsert it.

The health care team faces difficult choices regarding Mrs. Smith's care. Foremost among them is whether her physicians should surgically insert a gastrostomy tube in spite of her aversive behavior. Mrs. Smith has never made a living will, nor has she indicated to family or friends at the nursing home what her preferences would be regarding life-sustaining care in this sort of circumstance. Both her son and daughter have stated that she would nevertheless not have wanted a gastrostomy tube inserted and would, if she could presently decide, prefer an earlier death to being sustained indefinitely in the twilight of her minimally functional condition. In defense of this claim, they note that she has always been a very active, independent person and that she avoided doctors whenever possible.[11]

The standard approach to deciding what ought to be done on behalf of incompetent patients, in law and in bioethics, invokes two substantive principles: a substituted judgment principle and a best interest principle. These principles never conflict because they are lexically ordered: the substituted judgment principle must be applied first, and the best interest principle is to be used only if the substituted judgment principle cannot produce a decision. So a person deciding about the care of an incompetent patient must ask, first of all, what that patient would want were he or she able to make a decision and to communicate that decision. Often, though, a patient will have expressed no values or beliefs that are applicable to the precise question that needs to be answered, and trying to extrapolate an answer from the patient's general values or views of life or from the patient's personality or character traits can yield only speculative, equivocal conclusions. That is the case with respect to Mrs. Smith.

Mrs. Smith has not completed an advance directive (or living will) and has not designated anyone as her proxy decision maker. Mrs. Smith's reported character traits—independence and aloofness from physicians—are compatible with a wide range of interpretations. Nor do her general attitudes, preferences, and tastes provide determinative guidance about the proposed treatment. As Arras notes, "It is one thing to have negative attitudes towards aggressive life support, but it is quite another to actually refuse it in your own case" (1988: 938). Mrs. Smith continually pulls out a nasogastric tube and resists efforts to reinsert it, but how to interpret that behavior also is unclear. It could signal a strong, persistent, and current desire to reject life-sustaining treatments, or it could be a reflex caused by irritation from the tube. Given these uncertainties, how the substituted judgment principle might apply to Mrs. Smith is merely speculative.

Moving to the backup principle, Arras tries to determine what would be in Mrs. Smith's best interest by using a "moral triangulation" approach that locates her on a continuum of incompetent patients. At one end of the continuum is Mrs. Jones, a patient in a persistent vegetative state (PVS), who has only brain stem activity. She has no awareness of the world or what is happening to her and is unable to feel pleasure or pain. Mrs. Jones's only interest seems to be the remote possibility of misdiagnosis. Because Mrs. Jones's lack of conscious experience renders her incapable of being harmed, she cannot be said to be excessively burdened by treatments or to be "better off dead." So a decision to cease treating Mrs. Jones would have to be based on the conclusion that she no longer is a "person" (Arras 1988: 940).

Patients at the other end of the continuum do have interests, however. Arras's example is Mr. Black, a 90-year-old marginally functional patient who is suspected to have colon cancer but who refuses a laparotomy to confirm the diagnosis. Mr. Black seems to be competent, but psychiatric examinations disclose profound deficits in his short-term memory and substantial confusion about his condition and his surroundings. He is described as "pleasantly demented." Because Mr. Black clearly is a "person" with interests that can be promoted or frustrated, decisions about his treatment would be based on a patient-centered best interest test (Arras 1988: 940).

Where does Mrs. Smith fall on this continuum? She is somewhere between Mrs. Jones and Mr. Black. Like Mrs. Jones, Mrs. Smith no longer has an enduring "self" and has been reduced to "a mere locus of transient sensations." Like Mr. Black, though, she possesses some conscious experience in her ability to feel pleasure and pain and to have perhaps rudimentary emotions. Given the similarities to Mr. Black, one could try to ascertain what is in Mrs. Smith's best interest. To justify termination of treatment according to the best interest test, two criteria have to be satisfied: (1) The burdens of a patient's life with the proposed treatment would clearly have to outweigh whatever benefits might be derived from continued life; and (2) Further treatment would have to be inhumane due to the presence of severe, uncontrollable pain. With respect to the first criterion, it is hard, as Arras points out, to discern any burdens of continued treatment for Mrs. Smith: "In the absence of any persistent and severe pain … it would appear highly doubtful that the burdens of Mrs. Smith's continued existence clearly and markedly outweigh the benefits, even when the benefits approach zero" (Arras 1988: 942). So the decision that follows from the best interest test is that the G-tube should be surgically implanted in Mrs. Smith and her life should be maintained indefinitely with artificial nutrition and hydration.

That result seems decidedly unpalatable because the best interest test focuses too narrowly on pain and ignores other important moral considerations such as privacy, dependency, dignity, and bodily integrity. Unfortunately, Mrs. Smith has no awareness of how her dignity might be despoiled or her privacy violated. Any appeal to such concerns would have to invoke Mrs. Smith's prior preferences by claiming that she would not have wanted to be so utterly and completely dependent; that she would not have wanted to have her privacy violated so intrusively; and that she would not have wanted to have her dignity so deeply insulted. But that argument represents a return to the substituted judgment test, for it is an attempt to figure out what Mrs. Smith would want were she presently able to decide for herself.

The attempt to apply the principles of substituted judgment and best interests is going in a circle: the inconclusiveness of the substituted judgment test leads to the best interest test, but dissatisfaction with its narrow focus leads back to the substituted judgment test. How does one escape that circle? Arras's proposal is to shift from substance to procedure. Usually, in situations involving incompetent patients neither the substituted judgment test nor the best interest test can be applied given the sparse, tenuous, and ambiguous evidence that is available and the diverse conclusions that caring, sensitive, and responsible people are likely to derive from that evidence. So the question that must be asked, Arras contends, is not *what* should be done, but *who* should decide, and that determination is central to the decision-making process. Arras concedes that there are clear cases of overtreatment and undertreatment of patients but points out that between these extremes there is "a vast gray area …where the patient's best interests will remain unclear and largely inscrutable" (Arras 1988: 943). In this morally nebulous realm, the key moral question becomes who should decide about the care of the patient, and Arras's answer is that knowledgeable, concerned, well-intentioned family members are in the best position to decide what is best for the patient.

Arras adds two qualifications, one of which is uncontroversial and one not. The uncontroversial qualification is that, as in the case of K'aila's parents, there is a substantive constraint on this decision making: "the decision of a trustworthy surrogate should prevail over objections from caregivers, unless the latter can show a clear violation of best interests." The controversial qualification concerns what kinds of considerations are morally relevant to surrogate decision making. Arras acknowledges that "many caring and well-meaning family members will also want to weigh the impact of continued treatment upon themselves and the family unit," and he endorses that position: "This is to be expected and should not give us grounds for concern so long as the case originally falls within the gray area of ethical ambiguity, and so long as the interests of family do not *clearly* violate the best interests of the patient" (Arras 1988: 943; emphasis in original).

Shifting from the application of substantive norms—what to decide— to the design of decision-making processes—how to decide—is not just practically helpful, however. It also is philosophically attractive because, as we will see, it protects bioethics from the arbitrariness and subjectivity that ensues when principles, whether substantive or procedural, cannot be applied to produce definitively correct answers. Arras's procedural norms suffer from the same indeterminacy as the substantive norms of applied ethics. For example, they cannot sharply circumscribe the boundaries of that capacious "gray area of ethical ambiguity"; cannot comprehensively identify which members of families are "involved" and "well intentioned"; and cannot precisely instruct families about when and how to consider their own interests as they make decisions about the care of an incompetent relative. As with substantive matters, procedural decision making often requires judgment. The deeper lesson of Arras's analysis is that, given the inevitability and prevalence of judgment, which is required whenever there is no simple answer, the rational alternative to inaction or arbitrary action is to design a right way to make judgments. That is what the women making decisions about whether to try to become pregnant after receiving genetic counseling did, that is what K'aila's mother did, and that is what the judge endorsed. That is also what is done outside of ethics, not only in law but everywhere that judgment is similarly required: in educating the young, in designing a new building or ornament, in solving crimes, and so on. In taking this approach we ground our ethics in a universal form of intelligent problem solving adequate to the tasks (see chapter 4).

2.4   The Scope of Morality and the Limits of Formal Bioethics

The principles of bioethics are to be applied to moral problems, but what makes a problem a moral problem, moreover, one to which the principles of bioethics are pertinent? Does the following vignette pose a moral problem, and if so, are the principles of bioethics relevant to it?

Vignette 2.5   Dr. P

Dr. P, who works as an internist in an inner-city neighborhood, is the physician for the 12-year-old grandson and the 9-year-old granddaughter of Ms. J. Ms. J's son is a homeless heroin addict. The grandchildren's mother, who was also a heroin addict, is dead. Because Ms. J, who is 79 years old, is becoming progressively more demented, it is not clear how much longer she can care for her grandson and granddaughter. Her son is also Dr. P's patient, but she is not. Ms. J participates in the Grandparents Who Care program at Dr. P's public health center, and Dr. P facilitated a support group to which Ms. J belonged. Dr. P also has done respite care for the family and has taken the children on several outings. Even though Ms. J is not officially her patient, Dr. P cares about her, feels an obligation toward her, and "treats" her in the context of a grandparent support group that Dr. P started because so many of her patients need psychological and material support as well as medical care. Dr. P agonizes about the boundaries of her medical and personal responsibility to Ms. J and her grandchildren, and she has even entertained the possibility of adopting the two grandchildren (Kaufman 1997: 11).

Many people would regard Dr. P's problems—or perhaps Dr. P's perception of her problems—as personal or psychological or emotional, but not moral. Yet the vignette poses fundamental medical-moral questions about who is a patient and what should be treated. These questions would not be recognized as moral questions relevant to medicine by those who see medicine as a separate scientific and professional enterprise; rather, Dr. P's worries would be taken simply to expose her misconceptions of the basic nature and goals of medicine. Nor would they be recognized as moral questions relevant to bioethics by those who equate bioethics to the application of principles. Moreover, Dr. P will be aware that she could not extend such care to every similarly needy family she might treat and that her self-assumed responsibilities are specific to her history of engagement with this particular family, and that context dependence would be taken to show that no moral issues are involved here because the norms and decisions of morality must be universalizable. But the questions Dr. P finds agonizing challenge the assumption that medicine and bioethics can be cleanly demarcated and segregated from particular, personal practice. They are questions about what medicine is, what a commitment to being a doctor means, and what the limits of a physician's responsibilities are. They are questions that blur the boundaries between the medical, the social, and the moral—questions in which clinical matters, family matters, and moral matters are intimately and inseparably entangled.

There is this to be said for recognizing Dr. P's struggle as a moral struggle. Dr. P is troubled in the same way that the women who received genetic counseling are troubled: she is trying to figure out how to live well. She is a physician whose professional role is central to her identity and her integrity as a person. She is questioning her role as a physician because being a good physician is, for her, crucial to being a good person.

Applied bioethics does not recognize her struggle as a moral problem because it does not conceive morality in terms of the good life, living well, and flourishing in particular circumstances. Applied ethics is a problem-solving tool and, as with any tool, it works only where the conditions are suited to its application. That requirement specifies and constrains the domain of bioethics problems. But why should such a truncated view of bioethics—and morality, more generally—be accepted, and does such a view capture why moral problems are so troubling and so momentous?

2.5 Conclusion

The vignettes in this chapter convey two crucial lessons about the nature of morality. One is that there is much more to morality than principles and rules. Principles and rules are an important component of morality, but far from all of morality. The other is that judgment is pervasive and inescapable in ethical decision making. Most of the time principles and rules, in concert with values, customs, and practices—all embedded in social, political, economic, religious, and legal contexts and in background world-views—guide our behavior with little or no thought; real morality is like the air we breathe. When moral problems arise, however, principles and rules, isolated from the complex, multiple contexts within which they are understood and operate, and with all their weaknesses, are of little help. The morally relevant description of a problem has to be constructed, the salience of principles and rules established, the scope of general terms and concepts determined, and conflicts between principles and rules resolved. Judgment is required for all of that, and once those judgments are made, the applied-ethics exercise of subsuming facts under norms is but a logical flourish. What the deliberations of the women after genetic counseling in chapter 1 reveal, and what the agonizing of K'aila's mother and Dr. P in this chapter reveal, is that the abstract norms of applied ethics miss what real moral problems are and what they mean to real people. Real moral problems force people to make choices about how to live well—and to die well—and about what kind of person to become and to be. People take those choices seriously, even if they do not consciously regard them as moral choices, and they deliberate about them seriously, albeit not in the manner prescribed by formal applied bioethics.

The core question for morality, then, is: how can people rationally deliberate about moral issues and make moral judgments about them? Applying principles or rules to deduce universally correct answers, as happens in logic and mathematics, our paradigms of rationality, is the hallmark of rationality in the Western analytic philosophical tradition. Because a main goal of moral philosophy is to establish the rationality of morality (or else threaten its death in arbitrariness), it naturally aspires to, and thus tries to emulate, formal reason. The consequence is that morality becomes applied ethics. But the endemic problems of applied ethics elude that kind of rationality—and in similar ways, ironically but instructively, as does the invention of logical and mathematical systems themselves. The attempt to avoid these problems by combining principles and rules into a moral theory that resembles mathematical theories also fails because judgment is inescapable in that endeavor too. Whence the philosophical enterprise of using formal reason to vindicate the rationality of ethics fails.

That outcome leaves two options. One is to maintain the equation of rationality with formal reason and concede the arbitrariness and subjectivity of the judgment that pervades not just morality but our lives. And the other alternative? We know that cognitively we are capable of dealing with moral problems, replete with all their entangled issues, context dependence, conflicts, and shades of meaning, because we see it done, in the women responding to genetic counseling and in ourselves from what we accomplish throughout our daily lives, for example, as we constructively resolve a dispute between our children, or juggle parental and work roles. The alternative then is to develop a more expansive conception of reason, along with its corresponding epistemology, that matches the basic cognitive processes involved and that can thus account for the rationality of judgment, in morality and in our lives. This latter, preferable option, and the ethics that accompanies it, is the project of this book.

3   The Problems Generalized, Diagnosed, and Reframed in a New Account of Reason

The widespread presumption in ethics is that moral reasoning is a matter of applying universal moral principles to the relevant descriptions of the particulars of situations and thereby deducing right decisions about what to do. The preceding chapters have presented the many practical and theoretical inadequacies in this depiction of ethics. What should the response to these serious defects be? Simply rejecting the model of applied ethics will not solve the problems. Doing that would leave the sources of the difficulties undiagnosed and provide no constructive alternative.

Anticipating, we believe that the twin errors from which the grip of applied ethics derives are mistaken conceptions of the fundamental natures of reason and of morality, and that we must dig deep to expose those root errors and use them to construct better alternatives. However, the abstractness of the issues, combined with the seeming obviousness of the conventional wisdom, make this task difficult. We begin by briefly examining the parallel situation in science, namely, the presumption that applying science to a situation is a matter of conjoining universal scientific principles to the relevant descriptions of the particulars of situations and deducing predictions and explanations from them. Our aim is to show how the same practical and theoretical inadequacies that exist in applied ethics are also inherent to this model of application in science. But in science their deeper origins are more clearly visible.

It is generally believed that ethics and science have little or nothing to do with one another because they pursue quite different goals—knowledge, in science, and goodness (or rightness or virtue) in ethics—and they employ quite different methods in doing so—logical principles and empirical observation are used in science; moral principles and normative intuitions in ethics. To the contrary, though, once the same root inadequacies have been exposed in both science and ethics, we can be confident that their source is a shared but inadequate conception of rationality, and once that conception is replaced by a more adequate one, the similarities between science and ethics will become clearer.

3.1   The Parallel Difficulties of Applied Science

For convenient comparison, scientific analogues of moral problems presented in the preceding chapters are given, followed by brief ethical parallels.

1. Problem: *the indeterminacy of what is relevant in a situation*. *Science*: the indeterminacy of what methods, principles, and models are predictively/ explanatorily relevant for a particular situation. In the case of research into the behavior of plasmas (very hot ionized gases), for instance, it is not initially clear whether standard statistical analyses are valid—perhaps free electrons are not randomized and so do not properly have a temperature— or whether wire probes (the crude but most usable method then available) with their interposing distorting sheaths would permit their sheath properties to be distinguished from plasma properties, and so on. In manifold similar ways each distinct situation raises anew the issue of what experimental methods, principles, approximations, boundary and initial conditions, and so on are relevant for understanding it. *Ethics*: each situation raises anew the issues of what features of the situation are morally relevant and what principles, values, situational characterizations, and so on are required to properly understand it ethically and so validly determine what ought to be done.

In each case these issues cannot be settled by formal considerations alone and are instead matters for sound nonformal judgment.

2. Problem: *the initial indeterminacy of which specific features of relevant factors best characterize a situation*. *Science*: the unfolding understanding of the predictive/ explanatory issues as the situation is investigated. Initially, for instance, plasmas were simply probed by single wires, but later multiple probes were developed that provided improved information, in turn allowing plasma temperatures and containment geometry to be chosen to obtain the clearest measurement conditions, and so on. This process of interlocking improvements is typical and central. *Ethics*: each new situation raises the issues of what is the most appropriate characterization of the problem, what methods should be used, and what kinds of solution would be acceptable. Coordinated understandings of each of these dimensions of an ethics problem unfolds as the situation is investigated.

Again, formal notions of reason like deduction and induction cannot, in principle, offer guidance.

3. Problem: *the incompatibility of principles*. *Science*: the incompatibility within and between theoretical and methodological principles. In the plasma case, for example, one can model the hot interior as either a fluid or a gas, and may model it thermodynamically or dynamically, but not two or more combinations of these at once. Wise scientists select now one, now another as appropriate to the current situation. *Ethics*: the incompatibility within and between theoretical and methodological principles. Deontological principles, for example, "do no harm" and "tell the truth," conflict with one another. Methodologically, deontological rule-following and utilitarian utility-maximizing produce conflicting outcomes, for example, between freedom and welfare over issues like vaccination, voting, and vagrancy.

When relevant premises for reasoning conflict, any deduction or induction can lead to arbitrary results and vitiation. What is required is judicious contextual selection of fewer consistent premises or modification of premises to construct a compromise among them (see chapter 6), but formal notions of reason cannot, in principle, guide that process.

4. Problem: *the incompatibility among basic values*. *Science*: the problem of incompatibility among basic epistemic values. In plasma research, there are situations where the dominant value is the refinement of probe empirical accuracy and others where the dominant value is valid testing of theoretical explanation (e.g., testing for the predicted occurrence of electron cyclotron resonance), but optimizing for each of these values requires distinct set-ups, modeling, procedures, and data treatments. Such conflicts can arise between any pair of the many epistemic values of science (e.g., among any pair of empirical adequacy, explanatory power, theoretical unification, or technological applicability). *Ethics*: the problem of incompatibility among basic ethical values. Ethically salient actions aim to promote values such as autonomy, integrity, equality, justice, welfare, happiness, pain-freeness, and esteem.[1] But attempting to pursue these values typically results in multiple conflicts (notoriously, increases in total welfare often come at the expense of decreases in equality, and vice versa).

This is another version of problem 3 and the same conclusion drawn in 3 above also applies here.

5. Problem: *the occurrence of deep, ill-defined problems*. Deep problems are those for which not only is the solution unknown, but the problem itself

is not well defined, so that the values that ought to drive its investigation and the valid methods to do so are unknown, unclear, or in dispute, as are the set of applicable theoretical models, the solution set, and the criteria for successful resolution. *Science*: deep problems abound wherever a sufficiently novel research domain raises anew the applicability of extant concepts, models, and methods. Astronomical investigation with Galileo's telescope was a deep problem because uncertainty surrounded how telescopes worked, what they could reveal, and so on (Feyerabend 1978). The same is true of plasma investigation with probes. Conversely, the emergence of particulars of data, theory, or method can transform conceptions of what the problem and the solution goal are, as did Galileo's observance of the moons of Jupiter and their phases. *Ethics*: deep problems abound in social life, for instance, ranging from how to throw a nice party for new immigrants, to how to regulate cloning.[2] And our conceptions of such problems can be transformed by the emergence of particulars of description, theory, or method, as when we discover that some deviant behavior by a teenager is motivated by boredom and not rebellion.

Formal inference is defeated by ill-definedness, and judicious nonformal judgments are required both to develop the missing information, infrastructure, and so on, and to understand their significance.

6. Problem: *context/situation-dependence and universality. Science* consists of concepts, theories, infrastructure, and methods that range in generality from universal (e.g., laws of dynamics) to highly specialized (e.g., genetically clean artificial insemination of monotreme ova) and everywhere between (e.g., Galileo's and electron microscopes and organic chemistry of the intracellular Krebs cycle). The more specialized support the more general while the more general guide the construction and validation of the more specialized. For example, an organic chemistry model of the intracellular Krebs cycle will employ some general chemical techniques such as determining assays, while also employing highly specific and specialized concepts and techniques (i.e., methods) concerned with intracellular chemical dynamics; it will rely both on general background knowledge drawn from physics, chemistry, engineering, statistics, and on methodology specialized to that context.[3] All claims are accepted into the body of scientific knowledge, but each on grounds appropriate to its class of valid contexts. Appropriately, scientists in flourishing sciences maintain many highly context-dependent specialized relationships with other scientists, often distant in expertise. Wise scientists aim for (feasibly achievable) generality but select and specialize activities, arguments, and relationships to suit their total context.

Ethics consists of concepts, theories, infrastructure, and methods that range in generality from (presumptively) universal (e.g., principles of justice and utilitarianism) to highly specialized (e.g., a woman deciding whether to try to get pregnant after having received genetic counseling) and everywhere between (e.g., third-party privacy walls for acquisition of sensitive medical information). The more specialized support the more general, while the more general guide the construction and validation of the more specialized. Flourishing is characterized and guided by the relationships, practices, and structures within families, communities, countries, societies, and cultures. Appropriately, those who pursue ethical understanding maintain many particular and highly context-dependent relationships with other ethicists, legal experts and scientists, often distant in expertise, as well as lay persons of diverse experience. Wise ethicists aim for (feasibly achievable) generality but select and specialize activities, arguments, and relationships to suit their total context.

Because rule-bound rationality ignores the profound dependence of flourishing on both situation and context—universal logical principles are necessarily independent of context—it is unable to contribute much to the process of flourishing.

Drawing together the many hints about the proper terms in which to discuss the natures of science and ethics that have emerged during the foregoing discussion, we briefly summarize the deep similarities between science and ethics, under four headings: problem solving, judgment, normative learning, and historical development.

Problem solving. All these parallels suggest that the conventional wisdom about the disparity of science and ethics needs to be re-thought. Both engage in normative pursuits to determine what ends are relevant and what ought to be done to pursue those ends. So they share the same overall procedural, problem-solving framework. Using this framework in either discipline involves the capacities of being able to recognize relevant problems and what is normatively important about them; recognize and assess methods for exploring opportunities for resolution; assess the outcomes; iteratively modify problem, method, and resolution accordingly; and the like, and do all that in context-sensitive ways that ensure the flourishing of both the normative endeavors embedded in, and the particular persons involved in, the institutional arrangements.

Judgment. Working within that common problem-solving framework and addressing the six general problems that science and ethics share (see above) requires making a crucial set of nonformal, skilled judgments of the kinds indicated in the preceding paragraph (expanded list is provided in section 4.1). Any adequate conception of reason has to encompass the rich variety of reasoning processes that we in fact use to intelligently recognize, characterize, and resolve our problems—from which it follows that we require a sufficiently rich set of reasoning skills, both substantive and procedural, to execute the foregoing tasks. It seems clear that those processes cannot be grounded in formal reason.[4] Judgment is a core human skill that is largely irreplaceable by rational formalisms. Judgment is rationally learnable and improvable like other skills, and, like all intelligent creatures, humans need abundant practical experience with contextual problem solving for judgments to be sufficiently relevant, attentive, perceptive, and judicious to support intelligent problem solving. This requirement clearly holds for individuals and is reflected in the intensity and length of practical training for any high-level skilled role, obvious in medicine and law where preparation involves both academic work and supervised experience; it should equally be reflected in training for all ethical decision making[5] as well as for research and for military or public service roles, in apprenticeships for skilled trades, and so on.

Normative learning. Equally telling for the role of judgment, even in learning about our norms, is that norms and normative methods have been massively elaborated over time in both science and ethics as experience has been gained, in a process characteristic of knowledge and skills that the human species acquires through learning. In science, Galileo, Newton, Bacon, and many others had to develop, systematize, and scrutinize elementary methods for obtaining empirical knowledge, even after 2,000 years of discussion of the idea. The historical fruits of these and subsequent labors are manifest in today's situation, where statistical inference, decision theory, and the like provide us with powerful, proliferated, and refined methodological tools. And norms governing acceptable procedures and outcomes/solutions have been greatly extended and elaborated: for example, in science into the 24+ array of epistemic values now recognized (see note 22). Nevertheless, no method has been found to apply universally unconstrained by other norms and methodological challenges remain as science develops, provoked, for example, by the deep and widespread consequences of the recent appearance of complex systems models across virtually all the sciences.[6] Similarly, contemporary bioethics has emerged from the developing practices of medicine and biology and has been spurred on by the aberrations of wartime research on human subjects and by profound social movements directed at eliminating discrimination and protecting rights. Simple principles and codes of ethics have been elaborated into complex amalgams of standards, guidelines, and policies, legal precedents and regulations, and case analyses—a process supported by university training programs, research institutes, and professional societies, conferences, and journals, just as it is in the institutional development of science.

As for ethics, then, consider how bioethics has recently undergone an important shift, from the narrowly focused ethics of doctor-patient relations considered in isolation, to an explosion of institutional ethics committees with wide memberships and broad institutional domains and the emergence of public-health ethics for policy formation and social interventions (e.g., for management of epidemics, vaccination programs, and the like). Of course, many problems will still be generated from the doctor-patient locus as well as from other institutional loci such as government-sponsored health programs. Correspondingly there has been an expansion of formal professional ethical and spiritual advisory roles to private firms as well as public institutions. Conversely, these institutional frameworks both facilitate and shape normative learning in the individuals who develop within them.[7] These are all invented social processes, creative constructions which we are learning about (as well as learning from) and elaborating as we go.

Historical development. Finally, the histories of science and ethics reveal key shared evaluative or normative features that pertain to an understanding of their natures. For each domain, several competing fundamental theories about its normative structure have emerged; for example, empiricist confirmation and Popperian refutation in science, and foundationalism and coherentism in ethics.[8] Yet each of the underlying simple models of reason and knowledge—for example, empiricism in science and rationalist foundationalism in ethics—fails to provide an adequate foundation. Consequently, alternatives have proliferated that, in various ways, reject being confined to narrow conceptions of reason and knowledge—for example, Feyerabendian and social constructivist accounts in science and casuistry and narrative accounts in ethics. In each domain a complexly interrelated set of irreducible values has emerged that guides the pursuit of problem resolutions, for example, empirical accuracy versus generalizability in science and equality versus need in ethics, the values tend to conflict with and exclude one another in practical situations. In short, in neither domain is there a single, satisfying normative conception of formal problem solving, let alone one that simply prescribes the application of principles or rules to problems to derive solutions. Yet in both science and ethics just this is what standard practice presumes.

The conditions for the resolution of deep problems typically are not given in advance but are discovered or constructed through the process of trying to resolve those problems. Experienced practitioners in science and ethics (and elsewhere) have learned how to do this kind of problem solving effectively. Their success is commonly attributed to good luck, serendipity, inexplicable revolutionary illumination, or mastery of an art. In fact, it is attributed to anything but a rational process of problem solving. Why? Because rationality is presumed to be restricted to formal rule-governed formulae for making decisions. But that narrow conception of rationality simply evinces the prejudice toward formal reason. To the contrary, the fundamentalness and ubiquity of deep problems, combined with widespread success in solving them, makes a strong case for inverting that bias and presuming instead that solving deep problems represents the pinnacle of our rational intelligence. Given that realization, the challenge is to develop a more expansive conception of reason that can support such a capacity and its achievements. The reward would be an account of reason and a correlative problem-solving process that would span the disparate domains of science and ethics, and hence have a claim to fundamentalness. It would also explain our proper conviction that it is more intelligent to invent useful rule systems and demonstrate their domains of validity and incompetence than merely to use them. Most important, it would open up much richer conceptions of problem solving in science and ethics that are better suited to the multidimensional, integrated problems that characterize our more complex societies. We present our first steps toward such a model and its philosophical framework in chapter 4.

To close this discussion we briefly consider the option of comparing the difficulties of formal applied ethics with those, not of science, but of social and political sciences, cultural theory, and anthropology. These more personal-social domains all seem closer to applied ethics as social practice than does scientific research, with its focus on empirical knowledge rather than social behavior and its regimented theoretical and experimental structures pursuant to that end. We agree that the more directly social domains are of especial concern to ethics, and that there is more sympathy to be found in the more social domains than is likely to be found in traditional philosophy of natural science for the kind of nonformal approach to reason that we pursue here. However, we offer three primary reasons for instead choosing science as our reference point. (1) Assumptions concerning the nature of reason run deep; often they are so deeply embedded as to be largely or wholly tacit, their users unaware of adopting them. In particular, assumptions about the role of logic are often considered so obvious as to require no defense. In this respect, the lack of contrast between ethics and the various social action domains, given their shared focus on persons and social action, would tend to obscure the underlying assumptions. That science is so evidently different from ethics makes it easier to compare the uses of reason in the two disciplines. (2) We have a much more explicit, precise, and detailed account of reason's roles in science than in any of the personal-social sciences. This makes it possible to more readily discern the similar account that must also underlie ethics. For instance, the formalist ethics paradigm of deduction of right action from principles and contextual information (in ethics) comes into sharper focus once we see its comparison with prediction and explanation (in science) as deduction from theoretical principles and contextual information. And in particular, we have a much more explicit, precise, and detailed account of formal reason's failures in the realm of science than we have in any of the personal-social sciences, making it easier to characterize a richer nonformal conception of reason and its applications. (3) The personal-social domains present us with large internal diversities and less of a tradition of deep, integrated development of positions with which to mount a proper comparison. This makes addressing the nature of reason in these domains a long, tortuous task, and one that has less cognitive force. For all these reasons, proceeding as we do is more practicable, forceful, and illuminating. Moreover, one of us (CH) has developed the view that all domains of problem solving share the same core cognitive process (see Hooker 2017a, 2017b, and briefly the cases in section 5.2). If that were so, it would undergird the comparison with science and provide a common framework for developing the correlative processes of nonformal reason and of deliberation and judgment formation.

3.2   Confronting the Dominant Analytic Tradition

The preceding discussion has been about setting up problems properly—a necessary precursor to solving them—and that discussion provides powerful motivation to pursue an alternative, more adequate conception of reason than its equation with logic or formal inference. But two substantial considerations militate against this endeavor: the seeming obviousness of logic as central to reason, and, equally, the seemingly self-evident nature of the Western conception of rationality (see below) that frames that preference. The former deems the project of seeking alternatives improper; the latter renders attractive alternatives wrongheaded, if not unthinkable. So, to complete our preparation for thinking afresh about reason, this section briefly summarizes the limitations of viewing formal calculi, and logic in particular, as the primary expression of reason and then considers the related limitations of the Western conception of rationality that underlies the focus on logic. The analysis is brief because the aim is not to provide an exhaustive critique of Western thought about rationality, but simply to convey the essence of its difficulties so that our positive development of a more expansive conception of reason can proceed. At the same time, though, the ensuing critique of formal reason provides valuable guidance for the construction of post-formal (i.e., nonformal) reason.

3.2.1 The Critique of Logic as Method

Formal logical reason assumes that rational procedure is confined to the use of valid formal arguments—i.e., arguments whose validity is determined by universal logical rules—where the use of those arguments to construct inferences is compelling (Descartes's clear and distinct ideas), not least because their normative power derives from their form alone. In deductive logic all and only valid arguments are those for which the equivalent argument form is a tautology, a proposition that is true by virtue of its form alone.[9] This strictly separates rationality as form from any content, elegantly entailing that reason cannot introduce any content errors of its own into the exercise of rational deliberation and decision making. That guarantee has been the irresistible attraction of formal reason for the certainty-oriented Western philosophical tradition.[10]

Logically valid arguments are indisputably part of rational procedure, and any account of rationality that cannot appropriately accommodate them is defective. The crucial issue, however, is whether logically valid arguments are the whole of rational procedure, or even indicative of the whole of rational procedure. It has seemed obviously so to many, so obvious that Boole described logic as the "laws of thought." Others have thought that the necessity which the premises of a valid argument convey to its conclusion is so strong that it must derive from a transcendental warrant rather than, as in science, from the support of successful practice. Moreover, since deductive logic is necessarily truth preserving, and truth is the chief goal of rational intelligence, logic is taken to be the rational-making feature of intelligence. The dominant logic-based philosophical tradition in both science and ethics has taken as its goal to show how each domain is rational because it conforms to logical reasoning. The idea that humans are rational animals because they are animals (additionally) equipped to reason logically and thus to pursue truth and practical advantage, has been part of the core Western tradition since Plato and Aristotle. It has been recently rejuvenated with a new plausibility and suggestion of universality by the metaphor of intelligence as computation. Practical rationality is limited, however, because the ends humans choose to pursue have to be declared nonrationally determined; but thereafter the means they choose to pursue those ends can be rationally determined according to their formal efficiency, for example, by cost-benefit maximization.

Yet despite its elegance and power, the conception of reason as formal just as obviously does not stand up to examination. First, because the invention of logics would not itself count as rational, removing rationality as central to intelligence. Second, because of all the roles it cannot fulfill which require nonformal judgment instead, noted in section 3.1 above. And third, because neither of the twin processes at the heart of scientific knowledge formation—induction and refutation—is fully capturable in formal logic.

Consider first scientific induction. The inductive process of arriving at a confirmed theory T on the basis of evidence E is well known to be equivalent to making a deductively invalid, hence illogical, argument: if T (a theory) is true, then, if initial conditions I obtain (e.g., those specifying an experimental setup), then empirical evidence E is true, and I and E are true, so T is true, which has the invalid deductive form "T entails (I implies E), and I and E, so T." Attempts to construct formal inductive logics to legitimize scientific inductive practice have been made, but despite the development of various formal calculi, three basic obstacles to such a theory of induction endure. (I) Formal form-alone inference cannot introduce any concepts not found in the premises, so any inductive inference to theory from observation will be excluded because theoretical terms are all unobservable so cannot appear among the data. (II) Inductive inferences cannot always be consistently conjoined to other inferences, deductive or inductive. Whence induction is irretrievably local, it has no capacity to generalize.[11] (III) Reasonable inductive practice does not support the form-content distinction required to form a logic because reasonable inductive form is a function of empirical knowledge.[12] So the first central intelligent process within science falls outside of formal logic.

Now consider refutation, the opposite experimental situation in which the evidence contradicts the theoretical prediction. The simplest logical expression of refutation is deductively valid: "T entails (I implies E), and not E, so not I or not T." The epistemic point of refutation, however, is not to record the mere appearance of a logical inconsistency and hence of an error *somewhere*, but to localize and thereby identify errors in order to eliminate them and thus improve knowledge. It is normally assumed that I and E are known because observed, but observation is notoriously fallible so a possible site of error. Logical inconsistency, the only resource available here to formal reasoning, cannot locate an error, let alone identify its source. So far as inconsistency goes, error lies in either I or T or E, or in any pair or in all three, i.e., anywhere is possible. Each option can perhaps be checked further, but if this search is done by uncovering inconsistencies, error localization evaporates into an endlessly proliferating cloud of possibilities. In sum, error identification and localization also must be reckoned an intelligent nonlogical procedure.[13]

Combining both considerations, logic cannot model revolutionary scientific transitions, such as that from Ptolemaic to Newtonian dynamics, because they creatively overthrow (cf. refutation) basic concepts and principles and replace them with improved ones (cf. confirmation), and that cannot be a purely logical process for all the reasons set out above. Revolutions are the resolutions of deep problems, and logic cannot resolve such problems. Notoriously, revolutions have ultimately been considered irrational for all these reasons, despite constituting our greatest cognitive achievements in science. That inappropriate outcome we can, and will, change (see chapter 4).

More difficulties could be added, but here we conclude by noting that science is most powerfully formulated mathematically, and mathematics has a history of (re)construction every bit as thoroughgoing and counterintuitive as that of science itself,[14] one that includes the development of all the different logics of recent times (modal, fuzzy, relevance, multivalued, quantum, …). How is that history rational? It cannot be a logical induction from experience without revisiting those problems, and it is no more plausibly a sequence of a posteriori recognitions of a priori truths than is science. In fact logic appears to have the same kind of fallible construction dynamics as science has. Contradicting this, since Plato, philosophy has been tempted to consider mathematics, including logic, a priori, an intuition of the necessary, part of a normative/factual dichotomy that runs across every domain and is intended to keep the normative uninfected by the empirical that Plato so distrusted. But in fact there has always been a strong interaction between normative and descriptive theories, which, both theoretically and practically, the sciences have forced and which that dichotomy makes incomprehensible.[15] It has not proven possible to defend this divide in any area.

In particular, it has not proven possible to isolate rationality theory from developing science because ultimately our theory of our rational capacities must be shown to be grounded in the actual operations of our embodied minds, to be possible developmentally through our actual educational/scientific practices, and to be biologically feasible. This interaction of the putatively transcendental with the factual invites a more naturalist approach in which theories of both normative and empirical subject matters are treated as fallible and as being learned and revised in essentially the same manner, and thus as capable of interactively developing in the ways we see them actually doing. In short: Any adequate conception of normative knowledge has to be able to encompass the rich variety of normative judgments and learning processes that we in fact use to intelligently frame and resolve our problems and thereby to improve our knowledge. When logic is thus viewed as a theory of reason to be judged against our actual best practices, it thoroughly fails to satisfy this principle.

This quick survey implicates the restriction of rationality to formal logic as the source of the difficulties with rational procedure in science and ethics. Recognizing the parallel difficulties in applied reasoning across two large areas of human life, science and ethics, should instill confidence in the broader generalization that the same difficulties occur wherever we attempt to apply general knowledge. A trial of situations in areas other than science and ethics will quickly confirm that universality for the reader. Conversely, the universality of the difficulties suggests that their root lies deeper than the particular natures of their domains, and that we should look for it not in the peculiar complexities of scientific, ethical, technical, or other forms of decision making but in the constrained, formal nature of the reasoning that underlies the rule-bound, deductive conception of applied knowledge.[16]

For many thinkers, though, it is difficult to even conceive of an alternative to rule-bound rationality: the assumptions of the dominant Western philosophical tradition about the nature of rationality are so deeply ingrained that they sometimes are invisible. Deconstructing that tradition and showing how there can be alternatives to it should help us to see beyond those assumptions.

3.2.2 Critique of the Western Philosophical Tradition

In antiquity the rise of philosophy reflected a distinctive opposition of reason to custom, religious commitment, and brute power, a commitment to a life founded on knowledge and practical expertise developed with and grounded in reason. For our purposes here the essential ideas of the Western intellectual culture that emerged are twofold: (1) humans are rational animals, and it is their reason that distinguishes them from other animals; and (2) the essential project of reason is transcendence of imperfection—ignorance, prejudice, bias, egocentrism, anthropocentrism, anthropomorphism, and the like. Imperfections are transcended through a process of reasoning that improves our state or condition with respect to our ideals (cognitive, moral, and aesthetic ideals, and perhaps others).

Transcendence typically is taken to be effected through contact with an external reality that involuntarily conveys true content. The involuntariness indicates that we are not fooling ourselves with anthropocentric and egocentric projections or wish fulfillment, distorting the truth with bias and prejudice, and so on.[17] For rationalists, contact is with some transcendent world of abstract truths or forms that delivers certain knowledge of the constitutive norms characterizing a domain and the substantive content in that domain. In ethics, for example, substantive ethics is conceived as a system of nonempirical principles and rules of conduct, or a set of virtues. For science, substantive knowledge is empirical and contingent, deriving from the involuntary outcomes of contact with the external material or internal phenomenal world. Metaethically, normative domains could be comprised of either rationalist necessary knowledge or nonrational collective conventionalism and/or individual subjectivism.[18]

Most important for our considerations here, what both rationalism and conventionalism or subjectivism share is the removal of normative knowledge from doubt and thus from rational scrutiny and learning. Rationalism renders the normative rationally self-certifying, and conventionalism and subjectivism exempt the normative from rational appraisal. This concordance reflects the influence of a third, equally venerable and pervasive principle: (3) the reasoning process itself must guarantee transcendence. There must be no possibility of the process failing when it is used correctly. Reason, that is, must defeat Cartesian or strong skepticism: doubt must be impossible. For that to happen reason must be constituted in a tool that satisfies three requirements: (R1) *Normativity*: It must be normative, i.e., capable of distinguishing legitimate understanding from misconceptions (truth from falsehood, right from wrong); (R2) *Imperfection-free*: It must not introduce imperfections into its operation; and (R3) *Transparency*: That conditions R1 and R2 are met must itself be beyond doubt, i.e., must be infallibly recognizable.[19] Within this framework, logic looks like the ideal expression of reason and, with formal maximum expected utility (MEU) for pragmatic decisions, like reason's only expression. So formalism seems to be the metarational ideal account, indeed essentially the only account, of rationality. These conclusions seem to follow because of the prima facie clear and intuitive way logic-based and MEU formalisms meet the three requirements for an adequate theory of reason.

Complementing this formalist framework, each formalist position is backed by a corresponding metaphilosophy. Central requirements for logical empiricism are: *priority*, the claim that metaphilosophical and philosophical doctrines belong to a First Philosophy that is logically and epistemologically prior to, and normative for, science (and hence all knowledge); *formality*, the requirement that all the major philosophical concepts and doctrines are ultimately reducible to specifications of logical structure; *foundational-ism*, the claim that there is an indubitable basis from which to construct knowledge (namely, observation and formal reason); and antiskeptical *certainty*, which is now seen to follow. Empiricists and rationalists fully share this metaphilosophy. Under this framework the ideal of reason is the state reached on (ideal) completion of the transcendence process, a condition of having a complete set of guaranteed truths. (This includes truth-tracking inference, since every valid argument is equivalent to a logical tautology.) Achieving the ideal of knowledge is also having a complete set of guaranteed truths. Thus the two ideals are equivalent. (This is no surprise because both are derived from logic; it is simply a sign that the issue of normative knowledge is removed from rational scrutiny under these conditions.) This formalist account of reason does provide a program for pursuing the ideal of reason. In fact, formalist assumptions are so deeply built in to traditional analytical philosophy that their acceptance too often goes unnoticed.[20]

However, given the multiple difficulties of the formalist approach noted above, another, richer program for specifying rationality is needed, one that will subsume formal reasoning as a small, but important, component.[21] And there is room to move outside the confines of the formalist options. For instance, between the rationalist and conventionalist/subjectivist poles lies a range of constructivist positions where reason plays its role in the reconstruction of normative ideals, as well as in direct problem solving and decision making, as part of the learning process. Developing this possibility places normative knowledge on a par with other knowledge. Although doing so rejects the transcendentalism of the dominant Western tradition, this option is not incoherent, and it is supported by everything we know about the historical development of normative systems and about the general ignorance and limitations of humans, and by an evolutionary perspective on the origins and fallibility of our intelligence. From this perspective, and following Karl Popper, the demand for certainty inherent in traditional skepticism must be rejected as inaccessible to finite, fallible creatures like ourselves, and with that are rejected content foundationalism as well as requirements R2 and R3 on a system of reason (above). (Eternal transcendental truths are, one is tempted to think, the last gasp of a prescientific world.) This fundamental reorientation rejects the dominant Western ideal of reason as obtaining complete, certain knowledge.

Popper provides an argued case for bold conjectures subjected to rigorous testing as superior scientific method and correlatively adopts maximal informativeness as an alternative epistemic value to guaranteed truth, in turn opening the way to accepting multiple salient epistemic values.[22] Popper abandons the requirements of epistemological foundations and certainty, as he must do given his critique of observation and his high-risk falsification method, substituting for them the requirement that epistemology explain how progress in understanding is possible (trivial for empiricism, crucial but substantive for fallibilism). This in turn prompts a double shift away from guaranteed truth as the direct, substantive ideal of science. The first shift is to a Kantian reconception of the ideal, where it is understood as inaccessible in itself but holds open the permanent possibility of improvement; it becomes a normative goal toward which progress can be partially and fallibly made and to which progress adds value. The second shift is, correlatively, to a multivalued conception of epistemic progress and, with values often in mutual conflict, measured in terms of partial improvements in many surrogate epistemic values.

All this provides a rich framework that permits and rationalizes context-dependent trade-off among values in local methodologies, enabling the surrogate structure to partially and fallibly evolve over time as method itself develops. Moreover, to combat the misleading attractiveness of the formalist tradition, it is essential to keep in mind that the negation of each of the metaphilosophical doctrines of that tradition flows naturally from the alternative finitist, fallibilist perspective, where all claims are treated as limited and fallible and so potentially improvable through learning. Such positions permit a naturalism that is antitranscendental and prodevelopmental in its fallibilist learning approach to knowledge, both normative and substantive, while retaining a strong sense of normativity, i.e., without degenerating into conventionalism.

This critique and its accompanying naturalist alternative do not require any objection to the centrality of reason to intelligence or the conception of reason as focusing on transcendence of imperfections, the first two core principles of the Western tradition. It accepts these principles but gives them a naturalist reading. Similarly, the principle that transcendence is effected through contact with an external reality is accepted, but now, that reality is the natural cosmos for acquisition of both normative and substantive knowledge and there is no empiricist requirement of infallible access. It is also important to be clear that formal procedures are not rejected out of hand; there are many situations in which it is eminently reasonable to use formal procedures and even many cases in which a refusal to use an available formalism would be irrational. The point is that rationality is neither exhaustively nor fundamentally captured by formal procedures.

We are required now to take a new look at the nature of reason, the ideal of reason, and the program for reason. So far we have garnered several fundamental keys to constructing a more expansive conception of reason. First, there are the negative outcomes of *rejecting*

- priority, formality, and foundationalism as metaphilosophical principles,
- strong skepticism and the requirements of perfection and transparency for constitutive tools of reason, thus
- the Western ideal of reason as complete guaranteed truth, and
- formalism as an adequate account of the nature of reason and knowledge.

Second, there are the positive outcomes of *accepting*

- the Western conception of reason as central to intelligence, with the goal of transcending imperfections,
- that rationality evinces a genuinely normative capacity, with the normative a fundamental feature of the natural world,
- a Kantian account of ideals, with surrogates for those ideals that are finitely and fallibly but feasibly pursuable,
- the interactive development of norms, methods, and empirical knowledge,
- the fundamentalness of constructive development where reason plays its role in the reconstruction of everything, including substantive ideals and its own ideal, and
- the fundamentalness of nonformal judgment and its skilled improvement as underpinning these constructive processes.

3.3   Naturalism Frames a Conception of Reason Suitable to the Pursuit of Science, Ethics, and Life in General

The framework for freeing up a richer conception of reason proceeds from a shift to a naturalism of finite, fallible agents.[23] Humans are part of the great evolutionary tapestry of life, differing in their capacities from other organisms by degree, not metaphysical kind, and often inferior in various ways (to eagles in sight, and so on). The improvement that reason provides is to be construed as a (partial) transcendence of imperfections within natural life, not a transcendence of natural life.

Most of the radical implications of naturalism for developing an account of rationality derive from two encompassing features of organisms described by science: fallibility and finitude. Section 3.2 has already noted the radical implications of thoroughgoing fallibilism, in particular, the explicit rejection of priority and foundationalism, the skeptical assumption, the requirements of perfection and transparency, and the formalist ideal of reason. Dually and constructively, fallibilism directly motivates (a) the adoption of a Kantian account of ideals with surrogates—but all fallibly learnable through the interactive production and improvement of norms, methods, and empirical knowledge—and (b) the fundamentalness of rational constructive development as in fact the learning process through which we pursue, and improve, our ideals and surrogates . Before examining the equally radical implications of finitude, it is useful to briefly review the general nature of a thoroughgoing naturalism.

Broadly, the term *naturalism* as we use it refers to the position that the whole of life forms a natural (but not necessarily simple) unity, i.e., a single natural order or natural system, however complexly its parts are internally interrelated.[24] An important immediate consequence is that only divisions supported by empirical evidence are to be introduced. Divisions imposed by human fiat are rejected; for instance, a priori metaphysical distinctions between mind and nature, linguistic and nonlinguistic intelligence, normativity and descriptiveness, and reason and cause are rejected. Naturalism is committed to treating humans as a natural species and hence to understanding human capacities, including the capacities for reason, knowledge, and ethics, as natural capacities that have evolved biologically from more basic capacities.[25] This commitment imposes obvious constraints on the acceptable construction of theory. In particular, although philosophical theories are more general than scientific theories, the construction of a philosophical theory is considered part of the general epistemic process of understanding our world. Thus philosophizing yields fallible theories that are on an epistemic par with scientific theories. Albeit more indirectly, they are justified by the same kinds of explanatory considerations, and, up to a reasonable critical reserve, they must cohere with science.[26]

Human beings, like all other creatures, are imperfect: ignorant, as a result of their finitude, and in error, as a result of both ignorance and fallibility. Moreover, these imperfections run deep. Humans are born ignorant, not only of what the world is like but of what their own capacities are, and ignorant of what knowing and being rational (and loving) are and of what their pursuit involves. Everything more sophisticated than elementary responses has to be learned.[27] Human error involves a wide variety of biases, for example, prejudice, egocentrism, anthropocentrism, and anthropomorphism, and pervasive inaccuracies, reflected in the failure of actions to satisfy goals and in inadequate beliefs and norms that motivate and direct inappropriate actions. In this condition human developmental tasks are central and urgent, but constrained and complicated by the pervasiveness of human finitude and fallibility. To be relevant, normative theories have to take the imperfections and limitations of human beings as central to what they prescribe. This realistic qualification applies, in particular, to the theory of rationality. As we shall see, however, this qualification does not preclude creating and striving for ideals. Rather, it requires that the standards that guide such striving must be attainable by the beings we are (since ought implies can). At the same time, those standards must also allow for cognitive processes capable of being guided by ideals and formulated in terms of improvement in the attainment of those ideals.

From this perspective reason pertains to the developmental process, broadly understood to include all constructive self-change throughout life: reason is the organizational capacity to efficaciously prosecute imperfection-reducing / improvement-enhancing development.[28] Rationality is located in the character of the improvement process itself. The formalist account of rationality, in contrast, locates rationality in the character of the end product produced (certain truth). In our adaptable, but finite and fallible, condition, however, we have no specific pregiven ideals as goals, and we are not equipped to recognize transcendental conditions like certain truth, so such a conception of reason is inaccessible to us. Reason regulates how we go, including pro tem constructions of our anticipated ends. Moreover, a focus on end-product in our conception of reason prevents recognition of learning about, and developing the structures of, rationality itself as we go, but this self-improving learning is crucial to being rational (see chapter 4). It is the reflection in philosophy of resolving deep problems, discussed earlier. What it is to be rational, or ethical, are deep problems that have to be clarified and reunderstood as we proceed, and similarly for all the questions that make up philosophy. Like the baby learning to walk, as our journey proceeds successfully, our control over and understanding of our going also improves.

Reason as logic purports to deliver variously guaranteed truths, and fallibilist naturalist accounts cannot because they focus instead on the (fallible) process. In fact, trying to characterize reason naturalistically in terms of its outcomes is an inherent mistake since ends can no more be foreseen than can theories, and it is self-defeating since, as Popper made clear, aiming at security (read end-product reliability) and at informativeness are in conflict, a security emphasis missing the deeper, riskier insights while too strong a focus on risk-taking neglects assessment of what we can reasonably rely on as a basis for further exploration. Practically, confining method to learning only from failures or to learning only from successes are equally too slow and too risky for learning; finite, fallible organisms learn best from mixtures of failure and success. Ultimately, its practical success is the only way we can naturalistically (but fallibly) judge the adequacy of either a proposed specific solution for a problem or a proposed rational procedure for solving a problem. Rational practices should normally confer improved success; if they do not do so, and given that errors in theory and data are reasonably shown to be absent, we have reason to doubt the rationality of the practices. Improvement in pursuit of other goals is not just the goal of reason; it also is the lever for improving reason itself.

But how is the success of a rational practice recognized? Like other features of developmental processes, success cannot be fully characterized in advance, only in retrospect. At any time what counts as local success is simply satisfying the values delivered by the operative norms. The architect of a bank branch will look for design proposals that increase the value to the bank, whether in terms of short-term profitability or occupational health and safety, or longer-term client reach, market share and visibility, scale of client enterprise, and so on. But goals might differ over the course of the problem solving, perhaps changing the values deemed relevant (e.g., as public services or "green" principles are incorporated into the bank design), or shifting their priorities (e.g., greenness becomes a more desirable goal). As long as such changes in goals and values occur as problem solving proceeds and they increase the ranges of possibilities explored, evaluated, and accumulated as successively rich guides to next steps, they form one of the chief indicators that a powerful problem-solving process is ongoing.[29] Since such changes often cannot be anticipated and might only emerge during the course of the problem-solving process, global or overall success in increasing the satisfaction of values cannot be specified in advance.

Naturalism is adopted because it is fundamental to our most successful way of understanding our world so far—science. The history of science since the Renaissance reveals that progress has been made by naturalizing what was previously taken to be nonnatural. From the rejection of the heavens as a nonnatural sphere of perfection, which accompanied the rise of modern mechanics, to the discovery that the heart is a blood pump rather than a nonnatural center of feeling, which accompanied the rise of modern medicine, scientific progress has occurred as prevailing anthropomorphic and anthropocentric projections of nonnatural agencies into the world have been displaced by understandings of ourselves and our world as part of a single natural order. A stubborn holdout is the insistence that naturalized normativity is not possible. The theory of rationality developed here displays the same naturalizing spirit that has brought success to science, yet it claims to exhibit a genuine normativity. That conception of normativity must, of course, represent normative claims as being as fallible and learnable as the empirical claims of science. But instead of being the unthinkable repudiation of the obvious transcendence of norms that traditional philosophy would claim, it is simply part of the ongoing naturalistic revision of our self-understanding as we achieve a better grasp of the nature of the world and of ourselves as beings in that world.

Hilary Putnam's attempt to argue that reason cannot be naturalized provides useful support for this approach.[30] Putnam proposes a set of four conditions or constraints on any acceptable account of reason and supposes that together they rule out any naturalized rationality. But any naturalist norm will satisfy these constraints, so we are free to adopt them, and not only for reason but also for ethics. The conditions are these: (C_1) Reason is irreducibly normative; it is neither eliminable nor reducible to science. (C_2) Reason is an autonomous capacity; it cannot be defined in terms of other normative capacities. (C_3) Reason is to be conceptualized as a regulative ideal among other ideals such as truth, an ideal that we are permanently exploring but that we never fully attain; the exploration of reason is open-ended. (C_4) At any given time, however, the specification of reason is context bound, the relevant contexts being those of our cultures, traditions, practices, and procedures. However, a naturalist account of any norm (a fortiori rationality) will take the form of a theory of the nature of the norm and of the distinctive normative role involved, a theory that will be fallible and revisable, the one that best captures our current, fallibly learned knowledge about that normative dimension and its pursuit. But it then directly follows that such a theory of any norm will satisfy C_1, C_2, and C_4. And a Kantian account of the theorized norm, already adopted here, satisfies C_3.[31]

What emerges from this discussion of naturalism and Putnam's arguments are the following lessons for developing a naturalistic account of rationality:

- It is necessary to develop a general naturalistic account of normativity, and of ideals in particular, together with a theory of the particular nature of reason, including the performance standards against which the rationality of natural agents is to be judged. This should be a fallibilist account that treats all these concepts as theoretical constructs in a fallible, learnable theory of our normative capacities.

- To complete the satisfaction of C_1 through C_4, reason should be modeled, not in terms of specifying the character of the products of rational process, but as an open-ended organizational capacity focused on the process of improvement, i.e., on supporting the pursuit of all ideals, including an ideal of reason.

These principles are reinforced by the consequences of finitude. Formalist rationality theory stipulates that, to be rational, agents must formulate beliefs and values that are globally consistent and closed under deduction, take all factors into account when deciding, and so on. Not surprisingly, real people do not and, more to the point, in principle cannot check all their beliefs and values for mutual consistency, or hold them all simultaneously accessible for decision making, or determine all their implications for belief and action, or, typically, even perform all the desirable actions they can deduce from their beliefs and values. These "in principle" limits are remarkably strong; for example, any formal system that is undecidable cannot systematically review all inferences, and so forth, and this covers nearly all formal systems (a few of the very simplest excepted); and while we manage thousands of beliefs, it is not possible to check them for consistency, since even 40 or so would require more than the life of the universe to check for consistency using truth tables.[32] Moreover, pragmatic versions of these constraints bite still more tightly; for example, whenever the anticipated costs of pursuing consistency and the like are higher than the expected benefits, it is irrational to pursue these formalist requirements, and such situations are found from facing charging lions to selling shares on the stock exchange, from resolving children's squabbles to resolving adults' marital conflicts. The general purport of these lessons is that, for any finite creature, satisfying formal requirements does not stand at the center of being rational; rather the judicious choice of smart heuristics does, where judiciousness is a matter of judgment informed by experience and acquired expertise, current evidence, and an appraisal of current circumstances, including idiosyncratic aspects. Think here again of the women's decision making after genetic counseling (chapter 1). As Christopher Cherniak rightly explains, novel circumstances "are not peculiar to exotic scientific research; every child's cognitive development involves a similar type of cutting edge of knowledge, where incomplete mastery of new domains ensures unpredictability. In these areas what counts as 'reasonable care' just cannot be well defined" (1986: 120). This common plight poses the core problem for intelligent procedure: how are we to solve problems in new situations or new domains where issues of what concepts, criteria, and procedures to use are themselves ill-defined and undetermined?[33]

Throughout this book we both inveigh against and utilize simplified models of human agents. Contrary to appearance, this is a coherent, if occasionally delicate, position to maintain. The kinds of simplification we inveigh against are those that obscure the foundations of things. In our case, these have to do with assuming that agents can achieve logical consistency, make all deductions from a body of premises, survey all premises to obtain those relevant to a decision, try out all relevant trial solutions to problems, and the like. All such aims are in principle unachievable by any finite agent, and their adoption serves only to obscure the actual basis on which finite human agents tackle problematic situations, including ethical problems.[34] The kind of simplification we practice in this work is to leave aside all those complications that clutter, but do not fundamentally change, the model of human agency being used, and so could be "added back in" on occasions, and in degrees, required to explain agent behavior.

These latter simplifications fall into two kinds. (1) Imperfections in the exercise of rationality: for example, biases and prejudices, illusions, delusions, and preclusions.[35] It would be nicer to be without these, and some can be corrected through training to some degree, but most are empirically recalcitrant. But we can rationally discover these imperfections and learn how to correct for them, from eyeglasses to "data cleaning," and in this very process exhibit the true power of our self-improving rational capacities. (2) Complexities of our inner emotional and cognitive development and self-regulation that mean that we often do not make fully conscious, or fully balanced, or fully pursued, or fully understood decisions.[36] There will be plenty of situations where such factors play potentially important roles, for example, where doctor-patient relationships have both medical and developmental dimensions. Nonetheless, for the sake of simplicity we assume hereafter that none of the imperfections and complexities make an important difference to our general account and our basic understandings of examples.

At this point, providing an adequate naturalist account of norms will likely be a pressing issue for some; however, its pursuit would also take us away from our focus on ethics, so we set it aside here. For interested readers the basic ideas have been laid out elsewhere.[37] We close by emphasizing the process nature of reason—the rationality ideal is that of an ideal rational process, namely, to increase without limit the acquiring of relevant observational data, the canvassing of relevant reasons, the creative construction of relevant structures and processes, the subjecting of all to critical appraisal, the improving of the skilled judgments involved in doing all this, and the improving of rational processes that doing these things affords.[38] This process ideal of reason satisfies the core components of the Western *tradition* of reason: the centrality of reason to intelligence and the focus of reason on improving imperfections. However, unlike the degenerate formalist ideal, it rejects the Western *ideal* of reason, and the skeptical assumption. Instead, it is the right kind of rationality ideal for finite, imperfect creatures. This leads naturally to a rationality performance standard for finite agents: that rational agents try hard enough to carry out the reasoning tasks salient to the situation and feasible within their salient limitations in those circumstances. Here "trying hard enough" might mean context-dependently making all those inferences and/or consistency checks, using all those heuristics, and so on, that jointly contribute to increasing the feasibly achievable net benefit. This, for example, is a good overall characterization of the behavior of the women after receiving genetic counseling, and although we do not know enough about each woman to say how far she satisfied the required performance standard, it properly leaves open the assessment of inadequacies that might remain.

3.4   Conclusion

When it comes to the Western metaphilosophical tradition concerning rationality and normative knowledge, we have (section 3.2.2), first, the negative outcomes of *rejecting*

- priority, formality, and foundationalism as metaphilosophical principles,
- strong skepticism and the requirements of perfection and transparency for constitutive tools of reason, thus
- the Western ideal of reason as complete guaranteed truth, and
- formalism (logicism) as an adequate account of the nature of reason and knowledge;

and, second, the positive outcomes of *accepting*

- the Western conception of reason as central to intelligence, with the goal of transcending imperfections,
- that rationality evinces a genuinely normative capacity as a fundamental feature of the natural world,
- a Kantian account of ideals, with surrogates for those ideals that are finitely and fallibly pursued,
- the interactive development of norms, methods, and empirical knowledge,

- the fundamentalness of constructive development where reason plays its role in the reconstruction of everything, including substantive ideals and its own ideal, and
- the fundamentalness of nonformal judgment and its skilled improvement as underpinning these constructive processes.

What emerges from section 3.3 are the following lessons for developing a naturalistic account of rationality:

- The need to develop a general naturalistic account of normativity, and of ideals in particular, together with a theory of the particular nature of reason. This should be a fallibilist account, developed in interaction with other knowledge, formal, empirical, and normative, not held aloof from it, that treats all these concepts as theoretical constructs in a fallible, learnable theory of our normative capacities.
- To complete the satisfaction of Putnam's conditions C_1 through C_4, reason should be modeled not in terms of specifying the character of the products of rational process, but as an open-ended organizational capacity focused on the process of improvement, i.e., on supporting the pursuit of all ideals, including an ideal of reason.

The normative performance requirements imposed on agents must, to satisfy ought-implies-can, be feasibly achievable by those agents. Although normative ideals such as the rationality ideal cannot be attained, the performance standards for those ideals must be attainable. The logic-based rationality conditions (i) fail to capture the essence of being rational, and so are inappropriate as an ideal of rationality, and (ii) fail to provide achievable conditions, and so are inappropriate for a performance standard.

- The primary challenge for intelligent procedure is how to solve deep problems where the nature of the problem and what concepts, criteria, and procedures to use to solve it are ill defined and in need of resolution. Formalism is inherently incapable of responding to this challenge, yet deep problems and their practically successful resolutions occur whenever any creature masters surviving and thriving in a new domain.
- The capacities needed to act rationally are substantially nonformal and constructive (a) because they involve not only choosing among rational strategies but choosing among context-shaping options, including the shaping of problems and their outcomes; and (b) more generally, because our norms, including those of rationality itself, and our specific strategies all have to be learned as an essential part of the process of improving ourselves and our circumstances.

The logic-based rationality standard locates the rationality of agents in capacities such as deductive closure and effective determination of unlimited consistency. Consistency, deductive closure, and the like are properties that a completed account of knowledge is expected to have. The antifinitude assumptions of logic-based rationality assimilate the criteria for ideal knowledge and the criteria for ideal rationality, and that collapse goes unnoticed, thus unargued, because both are assumed to be end products. In its complete deference to outcomes that are necessarily correct, hence rational, logic-based rationality ignores the key structural features of the rational processes of finite agents listed in the lessons above. Whereas requiring consistency, for instance, is legitimate on all occasions (e.g., even when facing a lion it is never good to conclude that one should run both east and west) and there will be contexts where its pursuit is the appropriate primary activity (e.g., in mathematical proof construction), the unconditional pursuit of consistency is not rational for finite agents, so it cannot be conceptually fundamental to explaining rational processes. Yet process, not product, is central to the rationality of finite agents.

Rational process regards the skilled making of judgments about matters such as choice of context and heuristics as fundamental to the rationality of finite agents. Because it has no way to account for the rationality of judgment, the logic-based performance standard entirely suppresses judgment. In doing so logic-based rationality theory hides the more expansive conception of reason required to understand the rationality of finite agents. The posited unlimited capacities of logic-based rationality do not elide just some fundamental attributes of the improvement process—for example, context-dependency and heuristics—but the entire process, leaving only a presumed end product. An important distinction also is lost—the distinction between a *process* that enables us to (partially) transcend, i.e., reduce, our imperfections, and the *ideals* toward which improvement moves us closer, for example, knowledge and truth.

With this preparatory understanding of the need for a more adequate, more realistic account of rationality, we now can turn to the task of developing a naturalistic account of reason for finite, fallible agents.

4 The Problems Resolved in a New Account of Reason

The discussions that culminate in chapter 3 are not purely critical; they also provide creative insights into the kind of account of rationality that is needed to rationalize the actual capacities for learning and problem solving that humans, with all their finitude and imperfections, in fact have. In this chapter we construct a broad, general account of rational processes; one that responds to the challenges and insights raised so far and that suffices for our purposes here, which concern ethics. Developing the relationship of the central core of this rationality with risk taking, reasoning in its various modes (e.g., counterfactual, probabilistic, modal, and causal), and the like, on the one hand, and, on the other, responding to the manifold imperfections in reasoning that cognitive science has in fact shown, are each large and proper undertakings, we acknowledge; but they are for another day.

4.1 Skilled Deliberative Judgment

The foundational capacity underwriting our rationality is the exercise of skilled deliberative judgment.[1] Every step in problem solving requires judgments. To illustrate this claim and motivate its adoption, consider the following essential steps in problem solving and the kinds of judgments they involve, in each case illustrated in the women seeking genetic counseling (chapter 1). (1) Establish the relevant normative and factual aspects of problems. The women refused the implied normative/descriptive division offered by the specialists and chose to include a much wider range of norms concerned with identity, family, and beyond. (2) Determine how these aspects are properly understood and assessed in tandem. For instance, the women needed to determine what was fair for the child given the child's potential health burdens and how they might cope as parents. (3) Resolve conflicts among norms or values through constructive compromise. The women needed to understand how securing family support might conflict with risking and raising a seriously affected child, and what compromises between them could be accepted. (4) Identify options, their accessibility conditions, and their potential contribution to solutions. In the case of the women, for instance, there were various levels of support (or resistance) to consider from partner and other children, siblings, and their families, social and for-profit services, and what costs (monetary and not) each might involve. (5) Identify relevant methods for pursuing options. The women's methods could range from determining what value has highest priority (e.g., with respect to meeting a child's physical needs), compromising among several values (e.g., among personal time caring, pursuit of income, and family indebtedness), and using MEU (e.g., when considering use of various purchased services). (6) Identify where problem, solution, and method require alteration to improve access to increasing value of solutions. The women's shift from providing their personal values to an MEU, to instead exploring postnatal scenarios of caring for themselves and their immediate and extended families represents a coordinated transformation of problem formulation, methods, data required, operative constraints, and solution formulation that provided access to richer, more insightful and more resilient solutions. And this short list of kinds of judgment barely scratches the surface of the judgments tacitly involved in our approach to even everyday problem solving; for example, nothing has been said about judgments of reliability (for instance, by the women of themselves post birth). None of these judgments, or any other judgments, can be replaced by formal methods. Judgments are fundamental to processes of deliberation.

Skilled judgment, which has to be learned and is fallible in its exercise, is valuable and ubiquitous. Although no judgment is guaranteed to be correct, it is far better, for example, to have a certified, experienced airline pilot assess the weather conditions for flying rather than a passenger. Fortunately, judgment can be improved systematically through learning and training. The method adopted here is that of making and improving skilled practical judgments, which Harold Brown uses in his examination of rationality (1988). It is this learnable improvement that makes skilled judgment an appropriate basic tool for rationality because our deepest rationality lies in it. And because, thanks to higher-order learning, improving skilled judgment is itself a skilled process open to learned improvement, rationality itself can be improved over time—we see such improvement in the increasingly powerful scientific method, in the spread of ethical resolution processes in institutions, and so on.

Accounting auditors, who are faced with making professional judgments about the treatment of tax, investment, risk, and a multitude of other commercial affairs, many with clear ethical implications, have found deliberative judgment to be central to their skills, and there are accounting institutes devoted to training to improve skilled judgment.[2] Indeed, one's current skilled judgment rarely suffices for rationality, except momentarily, because current knowledge, acquired through use of those skills, is always feeding back to improve the skills themselves and to expand the situations in which skills can be applied. In the case of sensory perception, for example, the investigation of perception has both revealed the huge array of biases, distortions, illusions, and the like to which our perceptual processes are subject and also, in intimate interrelation with that, produced important improvements in, and corrections to, its functioning and a vast instrumented expansion of its capacities (microscopes—optical and electron; telescopes—radio, optical, X-ray; etc.). It is in this respect a useful model for the whole development of our understanding of rationality. Thus, it is rational to improve substantive judgments through rationality, and at the same time to improve the skilled procedural judgments that constitute the exercise of rationality.

While everyone can, and should, improve their skills, finitude and diversity of life situations will mean that inequities of skill and experience have to be recognized as an important and ubiquitous feature of life. Rational decision making is then often improved by having a variety of other experienced persons (both lay and expert) play appropriate modulating roles in making decisions, from ethics committees to governments.[3] This introduces in a natural way a necessary social component to being rational. It becomes obvious in the case of science, with its elaborated social processes of disciplines, research teams, and peer review, and is nowadays becoming equally obvious in ethics, with its elaborated processes in corporate and medical ethics, legal proceedings, governmental commissions, and so on.

For the traditional approach of logic-based rationality theory, the situation is reversed: reason eliminates judgment in order to secure certainty of truth. This suppresses both its individual and social roles. Of course, none of the procedural issues just outlined go away; they are simply suppressed by assumption of agent infallible perception, unlimited logical powers, and so on. These assumptions unfortunately deflect from acceptance of two fundamental agent capacities—agent finitude and fallibility—which so limit accomplishment. Finitude and fallibility also limit the ability that rescues us from those limitations, namely the capacity for higher-order learning while engaged in problem solving. Humans have nonetheless a distinctive history of successful problem solving, and this despite the problem solving processes resisting attempts at reduction to rules.[4] What counts as a good next jazz improvisation/ surgical procedure/ chess move/ research design/ medical care decision, for example, depends on what the previous improvisations/ patient history/ chess moves/ researches/ medical responses were and cannot be generated by any manageably usable rule set.[5,6] This situation comports best with skilled judgment processes, that is, with deliberation, as the expression of the basic character of rationality.

At this point an elephant in the room must be identified and stared down. The foregoing discussion develops a case that humans do in fact rely on skilled judgment throughout their lives as the basis for their rational capacities, but it does not tell us how that reliance could be justified epistemically. If the response is that they do so on the basis of skilled judgment, it looks circular, with appeal to arbitrary subjective conviction lurking in the background as its dismal fate. Yet before we despair, note that the situation is in fact the same for judgment about formal systems. Appeal to intuitive apprehension of a priori truths and the like is not only beyond naturalist reach, it is in the same difficulty as the appeal to judgment: there is nothing stronger to safeguard our judgments from error than more judgment. Thought otherwise should be arrested by the following. Incisive Descartes had no better answer than the notion of ideas that were so clear and distinct we just could not be mistaken about them, but categorizing which ideas are sufficiently clear and distinct requires a set of judgments. This is typical of the fate of all such proposals. Nor have we even always been right in our judgments, even in logic.[7] In practice correctness is judged by weight of judgment, from expert to crowd. This allows for social and historical critical evaluation, slowly rooting out errors—in fact by the very same means to be elucidated below for judgments. The power of these means should not be underestimated; they have allowed us to discover and develop powerful rational tools, from logics and mathematics to learning theory, perception theory, all manner of informing instruments, from glasses to microscopes, lathes to lasers, and more, and to correct all manner of imperfections in these in bringing them to their present, fallible, but mature forms. In short, we do not have, and in practice have not needed, a more surely founded basis for rational agreement than what we have for skilled judgments.

4.2   Four Resources for Rationality

In the exercise of reason, there are four principal resources or means for the improvement of skilled judgment: observation, the use of both formal and nonformal reasoning procedures, constrained but creative construction, and systematic critical appraisal. These four bundles of processes are utilized both by individuals and by communal groupings of various kinds: in science by individual scientists, laboratory teams, learned societies, collaborative and discussion groups, universities, etc., and more generally by the lay public at large to various degrees in various domains. And in ethics similarly by ethicists, formal institutional ethics committees, and less formal procedures and training committees and groups, universities, etc., and more generally by the lay public at large to various degrees in various domains. Each of these four judgment-improving resources (observation, etc.) generates a set of constraints on rational procedure, and these constraints mutually interact in complex ways. For instance, the requirement for systematic critical appraisal may lead to the introduction of new persons to an evaluation process (scientific or ethical), persons who synergistically also add to the variety of observations brought to bear on the situation. Together these constraints constitute both *dis*abling constraints (ruling out some possible actions as not rational) and also *en*abling constraints (making some actions rational that were hitherto inaccessible or inappropriate), and together these two sets of constraints structure and guide rational processes. In these ways, each of these means of improving judgment will contribute to both the elaboration of reasons for, and critical challenges to, our judgments—leading to improved judgments, for example, about the formulation of a problem, the method to be used to tackle it, and the proposed trial solutions to be pursued. And this process will also play various roles in developing our rational skills further. That they are the only kinds of judgment-improving resources available is itself a fallible judgment subject to reexamination as our understanding of reason improves. To elucidate the operation of the four judgment-improving resources, we briefly explore here how these resources severally contribute to rational processes.

4.2.1 Observation

Entering a room quickly, a person notices an unexpected movement and sees it as a black cat jumping off a chair, only to find upon closer inspection that it was a black scarf blowing off the chair in the draft from the opening door. The initial perceptual judgment was wrong, but notice these points: (1) without observation no useful specific judgments about the room could be made; (2) the perceptual judgment was not wholly wrong—there was a movement of something—and that judgment provided a useful starting point for further enquiry; (3) the error in the initial perceptual judgment can be corrected by further, relevant observation; (4) organized individual observation can be more effectively used to improve perceptual judgments about what is going on in the room than can disorganized or solely individual observation. This latter is because organized observation can focus the search on: (a) acquiring missing relevant perceptual judgments (can the object be located again?), (b) rechecking previously acquired perceptual judgments (does it still look like a cat?), and (c) arranging for the most suitable perceptual judgment-forming conditions (obtain more light to see properly? have other observers look from different locations?). These points extend to all observations.

In particular, they extend to ethically salient observation. As much as for judging their medical needs, medical practitioners learn, or should learn, to observe their patients in order to judge their ethically relevant needs. For instance, they observe to discover if a patient is in denial, or trying to avoid responsibility, or instead anxious to be told everything so as to feel more in control, or concerned about managing a difficult family situation. An important part of becoming an ethically experienced practitioner is developing such observational skills so as to treat each patient during the course of consultation with care for their overall well-being as they manage their illness in their particular situation. Especially given the sometimes deep interrelations between a patient's clinically relevant condition and prognosis and their condition as it pertains to overall well-being, ethically relevant skilled observation is not only inherently appropriate but often an important component of best clinical practice. In short, observation processes can be organized to improve the accuracy of perceptual judgment, and that provides one means of transcending subjective points of view.

Further, socially organized observation can be used to distinctively, and typically more powerfully, improve individual perceptual judgments because socially organized observation (1) can implement observation processes that are inherently unavailable to a single individual, for example, simultaneous observation from distinct locations, (2) can be more discriminating than individual perceptual processes, either by including more specifically discriminating individuals and/or by focusing more observational resources on a problem than any individual could provide, for example, a large search party, and (3) is more able to correct errors of perceptual judgment by using agreement among observers and rechecking as required. Moreover, socially organized training of individuals in observation can improve the power and accuracy of individual observation, as when an artist or a scientist is trained to observe the subtle relevancies that escape the untrained eye. The result is a virtuous cycle in which both individual perceptual judgments and social organization of judgment making are mutually improved, along with accumulating understanding of how to go about making still further improvements. All of these processes are intensely combined in the regulation of perceptual judgments in science.

All of them are equally available to support ethical judgment. Clearly, for instance, socially well-organized observation can improve ethical decision making by illuminating more aspects of situations and correcting individual observer biases, and the methods for arranging this can themselves be learned and improved. Child care workers concerned with ethical choices about parenting learn both what to observe about families in order to assess their parenting difficulties and options and when to call in others (a mentor, or psychologist) to check or supplement their observing. One of the side benefits of a well-functioning institutional ethics committee process is that it can make such resources available in a socially organized way in medical care settings.

The unique property of sensory observation is that it brings us into intransigent contact with the external world in a way that is sufficiently independent of current beliefs and practices to be epistemically useful. In particular, it is capable of robust agreement or conflict with current theoretical expectations, including those concerning its own explanation. And it is basic: absent some transcendental insight, physical interaction with our environment (plus the internal equivalent in kinesthetic and like signals) constitutes our only intransigent access to our world, so its epistemically useful functioning is a sine qua non for both science and ethics. (This much is correct in the traditional views: observation plays a special role in science and in the factual characterization of ethical situations.)

The job of rational experimental science is to increase continually the variety of useful observational interactions (e.g., by adding ultrasound probing to visual looking), their reach across kinds of situations and states of nature (e.g., telescopes extend our seeing into distant space and past cosmic states), their sensitivity to informative signals (e.g., neural probes can now detect the electrical activity of individual neurons), the precision of their outcomes (e.g., compare micrometers with rulers in length measurements), the range of their detection (e.g., a microphone can detect sounds both too high and too low in frequency for humans to hear), and their functional reliability (e.g., a spectrometer is a more reliable detector of light frequency than is unaided human vision). The underlying generic form of this formula is one that will apply to improvements in all four judgment-improving resource domains. In return, improved observation provides for improved scientific theories, including improved scientific methods. This is a core part of the upward spiral of mutually reinforcing scientific improvement.

That improvement also contributes to improved ethical informedness. It is easy to forget, for instance, that perception of distress and denial is an ability as much in need of improvement as any other observation of states, and that it is a complex procedural skill. And, as with observations in science, such improvements feed directly into improved ethical treatment and also feed back indirectly into improved observation via new standards of assessment and the addition of new aspects to be observed (e.g., of dietary regime or of family resources for dealing with conflict). The job of rational applied ethics is to increase continually the variety of these useful interactions (e.g., by adding conversational protocols to passive observation); their reach across kinds of situations and states of nature (e.g., adding observation in familial settings to that used in clinical settings); their sensitivity to informative signals (e.g., the verbal cues that a patient is distressed or feels excluded from participation); the precision of their outcomes (e.g., the many subtle ways in which behavior can indicate pain in dementia patients who can no longer communicate verbally); the range of their detection (e.g., game playing of various kinds can provide clues to disturbed states in young children); and their functional reliability (e.g., agreement between a medical practitioner and a social worker may be a more reliable detector of patient depression or anxiety than either's unaided observation).

Perception brings two additional advantages with it: (1) Perceptual interaction takes place all our waking time no matter what we are doing (and some important processes continue during sleep). While this has obvious survival advantage, an epistemic advantage is that we may use perception in a significantly undirected, curiosity-satisfying mode, just to "find out what is there." Whatever its ultimate rewards, curiosity is a core motivation driving our exercise of rationality in both science and ethics, for example, in exploring new data domains and creating new instruments of observation and in critically assessing trial outcomes from various angles. (2) Perception also has the epistemic advantage of generating conceptual content. Our creative intellects are dominated by the use of metaphor, and the primary source of metaphors is perceptual experience of interaction.[8]

But while individual perception is fundamental, it is but one of many physical interactions that can be utilized to construct information about the state and behavior of some system. Instruments of scientific observation and conversational protocols are other examples. All these interaction processes share the same basic features. (a) They are limited to a particular interaction modality, such as sound waves, and each has a characteristic frequency range within which it can register changes and beyond which it cannot (e.g., microphones extend human hearing, short interviews of children elicit higher quality of information concerning childhood trauma). (b) Being dynamical physical processes, they yield outputs to which multiple interactions, both salient and extraneous signals, distorting processes, insensitivities, malfunctions, and so on may contribute. (c) They transform their basic physical output into terms relevant for the larger investigatory process within which they operate. Natural perception involves a brain process akin to theory construction; in many ways the outcome perception is a plausible construction from our sensory stimulation, couched in concepts suitable for subsequent (inter)action (recall the earlier black cat/scarf example). Scientific and ethical observations are similarly constructed in terms of the expected theoretical conception of the situation in which the interaction takes place, whether what's observed is a high-energy particle collision or a patient's emotional response. (d) The observation performed through any of these processes (scientific or ethical) is subject to improvement by one or more of the following methods: (i) adding prostheses that correct or extend perception (e.g., lens to correct astigmatism, interview protocol to collect systematic data, etc.), (ii) specifying transformations of the raw data to represent particular features (e.g., angular momentum, depression) and to extract the relevant information from noise and bias (e.g., various statistical correlation analyses), and (iii) specifying the conditions for reliable observation (e.g., for humans, good light and no distractions, sympathetic and nondictatorial audience; for instruments, temperature range, incident vibrational frequencies and amplitudes).

In consequence, no observation process delivers simple, direct truths about the external world. To be epistemically useful, what observation yields must be subject to manifold critical assessments (judgments) concerning all of the above aspects. In other words, the data have to be interpreted.[9] Each of the complex theoretical and practical factors making up the observing interaction offers a corresponding locus of potential investigation, for instance, in responding to an observation that is judged anomalous (i.e., that conflicts with other observations or with a well-supported expectation). Conversely, though, each of these same factors is supported by whichever successful theories and practices represent the current scientific understanding of it (whether physical or psychological); and they in turn have multiple further connections. For example, for scientific visual observation the further connections extend from practical lens design through neuropsychology to theoretical statistics, and for ethically salient observation they extend through psychology to sociology and through interview protocol design to theoretical statistics. All of this together determines the potential range of epistemic ramifications to supposing the anomaly to originate with the observation and their plausibility as the origin of an error; that is, it provides an array of plausible explanations of an anomalous observation. That richly structured, multidimensional array is the basis of an interpretational latitude to observation for that situation; it remains undecided, with varying degrees of plausibility, whether observations are accurate or in various kinds of error, and there is the latitude to appeal to any one of these options if they would best fit with the results of other critical examinations.

That set of open possibilities is set inside another, broader set concerned with wider potential inadequacies. For instance, even if all its component processes operate correctly in themselves, observation may be so constrained by practices and expectations derived from beyond the observing context that it is incapable of uncovering the salient information—for example, because it is not undertaken in the relevant conditions to constitute a crucial test, or its output is incompletely processed, or it does not interact with the relevant features. Notoriously, for example, Galileo struggled to convince learned as well as lay people of his time that they could not observe absolute rest and motion, but only motions of one object relative to another. All the while he understood that unless the relativity of observed motion was accepted, simple observation would be taken to falsify his hypothesis of the motion of the earth around the sun instead of the reverse. When genetic specialists confine their observations of their patients to their family history of medical disorders and the results of any relevant laboratory tests, supposing this to suffice for understanding their rational decisions (see chapter 1), they have so blinkered their observations as to render them inadequate to their patients' needs and, perhaps unwittingly, also made it impossible for either themselves or their patients to understand any alternative for their counseling roles or the use of broader factors in their decision making.

The resolution of these latitudes in any given case is guided by method, theory, and practical skill, and by features of the specific context, all of which are themselves corrected and confirmed by refined observation. Here finitude requires that investigations be discontinued at the point where it is reasonably judged that the most salient issues have been sufficiently addressed for the moment.[10] It is the business of our other rational tools for improving our judgments—reasoning, creative construction and critical appraisal—to try, along with the constraints issuing from observation, to have science and ethical practice progress while keeping our interpretational options within manageable proportions. So, while epistemically workable, the relation between the claims of science or practical ethics and observation is complex and nowhere epistemically certain in the manner empiricism hoped. Settled observations are better understood as the outcomes of a complex whole-of-science or whole-of-ethical-evaluation process, even while observation is also the basic informational resource for those processes. In this double manner we improve our own rational means of knowing the world, and, most important, we simultaneously improve our resources for making further improvements.

We need to think of both science and ethics in these respects as dynamic systems, transforming their observing skills and capacities, and their use, as they develop. Judgment is required because of the absence of rules that are sufficient to decide such matters as how to develop an instrument and test its range, how to use it in the current circumstances, and how to interpret its output, not to mention broader matters such as whether to carry out an experiment or ethical intervention, when the experiment/intervention ends, and what response should be made to the outcome. Judicious judgment is required because all these aspects require coordinated critical assessment, in the face of finitude constraints. And expertise is required because only those expertly trained in the relevant disciplines and practices can make judgments judiciously on most of the issues.[11]

Like any other complex tool, observation can lead us astray as well as toward improved knowledge, ethics, and rationality. The roots of error lie within the design of each instrument as well as within our own limitations and imperfections. But, as our historical experience shows, this unavoidable risk does not prevent the development and use of improved tools and ought not to be allowed to suppress continued improvement. For when all of the specialized human and material resources are appropriately assembled (a matter of further judicious judgment), we generate the grand virtuous spirals of science and ethics, each ascending toward its ideal, improvements in each of observation, method, theory, and practice leading to improvements in the others, each step along this spiral constituted by a set of judgments, while each consequent improvement is used to improve our capacity for good judgment.

This process itself will conform to reason if at each step the judgments that tentatively close discussion are reversible and all of the elements are in principle open to criticism and testing against alternative interpretations and observations, and if in each case these resources for reassessment have been applied to the maximum extent judged reasonable in the circumstances. In particular, and to repeat, the rational tasks of each of experimental science and practical ethics are to increase continually the variety of these useful observing interactions, their reach across kinds of situations and states of nature, their sensitivity to informative signals, the precision of their outcomes, the range of their detection, and their functional reliability; and, to the maximum extent feasible, to bring the resulting observational resources to bear on the improvement of methods, theories, and practices.

4.2.2 Reasoning Methods

Ignaz Semmelweis, a nineteenth-century physician, was concerned, both clinically and morally, about the high rates of death from puerperal (or childbed) fever in maternity wards, and particularly concerned that the newest, most medically advanced ward had a higher death rate than did older wards. He surmised that contamination with "cadaveric matter" may cause childbed fever and reasoned that if this was the primary cause then there should be a significant difference between women served by doctors with cadaver-contaminated hands compared with the patients of doctors who had sufficiently cleansed their hands. An experiment to test this prediction resulted in decreased deaths among patients in the hand-cleansing arm. But Semmelweis's methods were either ignored by, or attracted general opposition from, the tradition-bound and/or self-serving doctors of his day. Semmelweis died at the age of 47, shortly after being committed to an insane asylum.[12]

Despite this unfortunate outcome, Semmelweis's reasoning from his causal hypothesis was effective and essentially correct: if A causes B, then if A is present, B will occur (contamination arm), and if A is the major or only cause of B, then the absence of A will result in a marked decrease in B (hygiene arm). Mill codified simple causal reasoning of this kind in what are known as Mill's methods; Semmelweis used what Mill called the "Method of Agreement and Difference."[13] The use of (generalized) Mill's methods lies behind much experimental design. This is an example of how methods of reasoning provide constraints on the set of judgments we can accept, both epistemic and, in consequence, moral—constraints that are just as powerful as those provided by the judgments that arise from observation. In the case of Semmelweis, his hypothesis was eventually borne out, and improved hygiene led to a large decrease in hospital maternity deaths.

The basket of formal reasoning tools of this sort includes at least formal logic, a codified set of rules for arguing validly, that is, in a way guaranteed to preserve truth; rules of causal reasoning (Mill); statistical analysis and inference; and risk-utility satisficing or maximization.[14] Each of these forms of reasoning adds valuable tools for improving judgment. To illustrate and appreciate this last remark, consider deductive logic. Its major advantages are as follows. (1) Localization: each deduction is decomposable into small, simple individual steps that can be independently checked. (2) Partition-ability: local reasoning about the valid deductive consequences of a small collection of propositions can be studied without reference to any wider set of beliefs and hypotheses to which they may be associated.[15] (3) Simplifi-cation: logic greatly simplifies information by restricting it to 1/0, yes/no, true/false judgments; all the graded, partly overlapping, multidimensional, complex distinctions the world may demand (e.g., that between oblate and spherical, culpability and responsibility) are replaced by 1/0 artificial divi-sions offering only approximations to real relationships. Simple judgments of this kind are also more easily checked for formal correctness. (4) Con-text independence: valid reasoning is a context-independent, simplifying procedure and is useful for communication across distant contexts where shared substantive assumptions cannot be made. For all these practical rea-sons logic is an attractive reasoning tool for strongly finite agents like us.[16]

In science and ethics logic can be used to test individual theories and models for self-consistency and to test pairs of theories/models for mutual consistency; it can be used to derive predictions/obligations, falsifications, and explanations; hence it can also be used to reveal assumptions, by showing what extra premises would minimally be needed to obtain a valid derivation. These aspects form a fundamental methodological skeleton for science and ethics. All these are examples of how the use of formal reason-ing increases the variety, reach, sensitivity, precision, and reliability of the arguments we can make in science and ethics, thus contributing to the development of both scientific and ethical theory/modeling and method. In these many ways, then, logical formal reasoning extends and refines our capacity for good scientific and ethical judgment. To this extent the tradi-tional account of the role of logic in reason is correct (even if the rationale is very different from the traditional one). But, in keeping with section 3.2.1, and particularly when applied to persons, with their concerns for historical and functional holistic integrity, it is equally obvious that it has limited applicability and needs to be enriched by the other forms of formal reasoning, as well as underlain by good nonformal judgment.

All formal reasoning methods also provide a means of transcending a local point of view, since everyone can use them and their conclusions are not user dependent. By widening and making systematic and public the demand for agreement about the consequences of methods, hypoth-eses, and practices, the use of formal reasoning again provides an important avenue for improving judgment in science and ethics. And, like observa-tion, reasoning can be socially organized so as to transcend individual judgment and improve method, in both science and ethics, through (a) implementing reasoning processes that are unavailable to most individuals, for example, complex computer programs, (b) being more discriminating than individual reasoning processes, for example, through employing many expertises, and (c) increasing creative deployment of argument and (d) critical scrutiny of claims and arguments through social communication and requiring suitable invariance of reasoning judgments across reasoners. All of these processes are intensely combined in the regulation of reasoning judgments in science, and can be just as intensely combined in ethical problem solving. As with observation, there is an ongoing reciprocal interaction between individuals and socially organized groups, in reasoning and improving reasoning, that can be powerfully mutually beneficial. Although when wrongly organized it can equally be damaging, requiring wise judgment to avoid, when productively organized the result is a virtuous cycle in which individual reasoning judgments and social organization of reasoning are mutually improved.

But despite its value, the contribution made by formal reasoning to the improvement of judgment should not be overestimated. It is, for example, inherently incapable of supporting reasoning from evidence or from principles beyond simple deduction (section 3.2.1) and simply passes uncertainty from premises to conclusions. Reason by itself, like perception by itself, markedly underdetermines rational belief. And further limitations could be added, for example, the noncreative nature of formal inference impoverishes formalist cognitive theory, since the capacity to construct new concepts, principles, methods, and practices is central to human intelligence. And there is no natural place in logic for heuristics, for methods that relinquish logical guarantees of reaching their goal in favor of being simpler and faster and still offering better-than-chance odds of being roughly correct (such as deciding on a tree to escape a charging lion); yet for finite, fallible agents these methods are central to rational behavior. Another limitation is that, ironically, the creation of logic itself is clearly a matter of intelligent judgment, not formal rule following, for all nondecidable fragments.

All this suggests that logic should be treated as a fallible theory, in principle like any other—a view supported by the deep relations between logic, algebra, and projective geometry that result in empirical dynamical theories exhibiting an underlying logical structure. This includes the nonclassical quantum logic, for which not all classical tautologies are quantum tautologies, suggesting that classical logic might actually be subtly false and at least that it has embedded empirical content. Logic might then be considered our most general theory of possibility.[17] A rational resolution of these issues is to be pursued by the very process that we find in the extant literature: open critical debate enhanced by further creative construction of all the relevant formalisms and the continued observational testing of the scientific theories in question (such as the practical use of fuzzy logic on the Tokyo subway and the imminent rise of quantum computation).

Time has been spent here on logic because it, more than other reasoning tools, needs situating in the larger space of naturalistic tools for improvement of judgment. After that examination, it might seem that the traditional alternative formalism for rational action, decision-theoretic utility maximization (or satisficing, for finite fallible agents[18]), is in a very different position. Utility maximization has no unbreakable rules (except that the maximization requirement be met); it is open to as many distinct values as there are recognized dimensions to utility, nor need those values be mutually coherent or immutable; and priorities among utilities can be highly context-dependent. Finally, and more radically, it allows the possibility that organized social practices may be emergent from multiagent interactions, including the systematic deliberative and behavioral practices of science, ethics, and law, to the extent that they exist. For these reasons, utility maximization offers a much more inclusive framework than logic for characterizing rational processes across many more kinds of situations in ways that provide satisfying explanations of capacities and behavior.

For instance, the scientist can be modeled as a risky investor in scientific credit, yielding a model of risky belief commitment that replaces inductive logic in a way that gives explicit recognition to the diversity of inductive situations, indeed that applies in both normal and revolutionary settings. In general, this framework makes the multitude of diverse scientific practices—both those that mutually conflict and those that are simply different—explicable as efficient ways of spreading risky cognitive investment across institutionalized disciplines. It illuminates the conflicting nature of multiple epistemic goals, and context-dependent trade-offs among them, by modeling them as utilities while providing for the explicit recognition of the context-dependent priorities assigned to them and their capacity to develop over time as sciences mature. It even allows epistemic rules to emerge from these institutionalized activities much as price does from market trading, thereby providing a new model for understanding their origin and historical character. For all these reasons the decision-theoretic framework makes deep sense of institutional structure as an intrinsic part of the cognitive capacity and coherence of science and makes the design of that institutional structure itself a rational choice, learnably improvable. A decision-theoretic framework, in contrast to a logic-based analytic one, encompasses such considerations naturally and thereby radically expands and enriches our conception of reason and its place in an understanding of intelligent action.[19]

But here too there are inherent limitations: maximization imposes huge demands on information and control of processes, so wherever these resources are scarce—for example, wherever there is deep uncertainty—it has no or little guidance to offer. For instance, it cannot handle situations where development of, and/or choice among, concepts, utilities, methods (e.g., investment rules) and decision contexts (e.g., games) play a central role in effective problem solving, because these actions either change the categorial structure of the decision problems or introduce radical uncertainty (or both).[20] Neither does maximization help to understand the nature and limits of practical forms of reasoning such as metaphor/analogy, casuistry, and narrative analysis. Finally, its very broadness reflects a lack of guiding normative structure: utilities are given no constraints on their structure; maximizations (and thus economic efficiencies) are as multiple as specifications of system, time frame, and utility structure. All this argues in favor of regarding MEU models as relevant only after a problem has been otherwise properly structured, where they serve in the dependent role of providing a criterion for ensuring (sufficient) efficiency (that is, waste avoidance) when executing the problem solution. MEU modeling is ultimately a formal device, and its shortcomings largely derive from that. Thus, however valuable some of its insights, utility maximization modeling is best treated as but one formal tool among many that contribute to rational improvement processes.

In sum, for both science and ethics a second crucial judgment-improving resource lies in our methods of reasoning: in both our formal methods—logic and mathematics, decision theory, and economics—and in our informal methods, including causal reasoning, reasoning by metaphor/analogy, casuistry, and narrative analysis. These reasoning methods severally and jointly assist us to increase the variety, reach, sensitivity, precision, range, and reliability of the deliberations we are capable of. Among other uses, we can use deliberations to improve our testing and evaluation of our methods, theories, and practices—including, in return, our theories and practices of sound argument.

4.2.3 Creative Construction

We have already seen an example of creative construction—the construction of reasoning formalisms like logic and mathematics—which at once makes clear the reality and importance of this category. Without the creative construction of reasoning formalisms, as the preceding section reveals, there would be no public frameworks for rational procedures, nothing to learn from concerning their limits or from which to attempt new creative expansions.

The point generalizes. Consider the practical example of developing paintbrushes with synthetic hairs.[21] Despite detailed copying of horse-hair bend and the like, initial attempts at synthetic brushes all failed, the brushes delivering paint unevenly (lumpily) and in a rush. Sharper scrutiny eventually made explicit a hidden, unquestioned assumption about what horsehair features were relevant: a paintbrush is essentially a spreader, assembling a vertical layer of paint in front of it that it spreads over the surface as it pushes it along, essentially in the way soil or sand can be spread by dragging or pushing an upright shovel across a surface. But suppose this assumption was wrong, and instead a paintbrush is better conceived as a pump, with the hairs storing paint in the interstices between them by capillary action so that pressing down on the hairs as the brush moves pumps the stored paint out onto the surface. In that case it would be most important to reproduce the structure of the hair interstices and their spread under pressure. Success soon followed.

This example well illustrates the centrality of creative construction—here a novel analogy—to problem solving. The analogical shift from spreading to pumping dramatically changes the fundamental suppositions about how the process proceeds and thereby redirects efforts to transfer the function to new materials, stimulating the development of new methods for interstice analysis, for hair design and construction, and for measurement of capillary actions, pumping pressures, etc. Analogical and, more widely, metaphorical thought is arguably the most fundamental mechanism of cognitive generalization, and we see how creative construction lies at its heart. It shows too that creative construction need not be explicit or primarily symbolically formulated (instead it might, as with pumping versus spreading, be grasped as practical motor behavior).

The example also illustrates the other fundamental feature of cognitively directed creative construction: the importance of the contextual constraints guiding appropriate construction. In the synthetic paintbrush case the chief general constraints on a successful change of analogy were that it should explain how paint flow occurs and on that basis lead both to insight into why earlier attempts failed and to detailed improvements in brush performance, and do so while continuing to meet functional constraints: resulting brush not too large, too heavy, too specialized, too expensive, and so on. In science and ethics generally, creative construction is constrained by at least five broad classes of contextual judgments, namely, concerning *relevance/salience* (for current problem, for "leading-edge" learning), *cognitive resources* (theory, methods, trial designs, data), *epistemic capacity* (performance requirements for a successful construction, specified as increases in epistemic values), *ethical constraints* (e.g., subject autonomy, public health, and safety), and *practical constraints* (financial, career, etc.). Sometimes access to superior solutions hinges on changing a constraint, for example, introducing futures trading to stock exchanges, which reduces producer risk while making markets more predictable for investors.

In both science and ethics these constraints give cognitive form to creative processes; they channel it, and so shape it, and thereby help to make the creative effort more feasible. Constructions have to be creative if something genuinely new is to emerge, but if the overall process is not also sufficiently broadly guided, the costs of more open exploration can quickly outweigh any value to be gained. Constraints are also essential to our capacity to assess a construction: a particular creative construction may be an expression of an individual's judgment, or that of a small group of individuals, but satisfying the constraints provides the means to test its quality in a manner that transcends the judgment of any individual or small group of individuals. Moreover, constructions are public entities and must ultimately be connected to public observations, arguments, and critical evaluations, bringing them under the other resources for rationality. In this stronger way, socially organized creative construction provides a way of transcending individual points of view while improving our knowledge and rational capacities.[22] Problem solving within strong but not determining constraints has provided much or all of our most profound creativity, from symphonic composition and jazz innovation through architecture to fundamental science, and ethics also, for example, in crafting compromise and allocating kidneys for transplantation (see, respectively, chapters 6 and 9). Overall, practical ethics shows a healthy vigor in adapting to the many new situations our changing world presents, from the ethics of genetic alteration and counseling through ethics of design to ethics of corporate governance, and this despite having to function largely in the public societal realm with its diversities and crosscurrents.

Creative construction, whether in science, ethics, or elsewhere, involves the creation of new concepts (e.g., transgender) and conceptual analyses (e.g., autonomy for the aged), new conceptualizations of problems (as in the case of the women who received genetic counseling), new analyses of assumptions (e.g., that autonomy is univocal), new theories or models (e.g., of doctor-patient relationships), new instruments and experimental procedures (e.g., novel interview instruments and protocols), new methods (e.g., the equally partial but equal best analysis of ethical policy options and their implementation[23]), new institutional roles (e.g., new peer review and ethics committee processes), and so on. Without these innovations most of the power of our cognitive processes is lost; science and ethics would be crippled. In both domains, every sound experiment, every sound technological development, every sound piece of modeling, requires myriad intelligent creative solutions to problems posed by awkward constraints, incompatible performance requirements, the need to discover unknown details of processes or performance limits of instruments, checking for undiscovered errors, and so on. Without an ability to find creative ways through these problems, our scientific and our ethical advances would have long since slowed to a crawl. Instead, they are accelerating under a myriad of multiplying, creative synergies among all these features and more: institutional creativity—new laboratory designs and new institutional ethics roles, new journals, new collaborative research arrangements, etc., all supported by technological creativity—new kinds of devices, new procedures, new quality control methods (including ethics); and so on.

Individual diversity in creative construction ensures a rich array of available constructions to consider, and that is essential to their productiveness of useful innovation, even while local sharing of expertises permits necessary cross-checking for errors and overlooked potential improvements. This applies to both science and ethics. In short, in both cases neither agreement nor difference is alone necessary or valuable; each is necessary and valuable in its proper role. A productive social organization has in fact a further subtle complexity: within their contextual constraints, creative constructions gain their distinctive normative powers from intricate, but systematic, patterns of difference and agreement across individual scientists or ethicists, for example, in ways that ensure that all the reasonably plausible alternatives for some problem solution are constructed and tested, along with selected "wild" constructions to test for present blind spots (as in the above example of early synthetic paintbrush hairs). In both cases constrained creative construction offers us new opportunities for organizing conceptualization, explanation, and investigation in ways that are sufficiently independent of current beliefs and practices as to provide deep or innovative learning. In the absence of transcendental insight, this has to be the way to open up our normative grasp, and to similarly transform our repertoire of physical interaction (including observation and communicative interaction) with our environment and each other and our methodological and explanatory resources. If these cannot expand synergistically with our expanding learning, then it is impossible to understand how finite, fallible creatures beginning in ignorance can ever improve their learning either in science or in ethics.[24]

Pressing the diversity parallel between science and ethics raises the more general issue of the ultimate nature and source of normative rational, epistemic, and ethical diversities within and between human groupings. For nonnaturalist accounts of norms as a priori, disagreement can only be due to some kind of ignorance that nonetheless gives way to a priori certainty under some conditions. By contrast, once naturalists distinguish between what is appropriate to finite, fallible agents here now and those conditions that may be specified in unachievable ideal states (e.g., complete consistency), it remains everywhere open to learn about the most adequate normative concepts, principles, and methods in the process of empirically applying reason, science, and ethics. And the parallels ramify. Compromise among norms is unavoidable and ubiquitous for finite agents, hence it is so in both science and ethics. Recognition of diverse local cultures and skills is equally unavoidable and necessary for finite agents, for without them we would lose a special diversity, that of those local traditions that develop the highly sophisticated, specialized skills requiring years of devoted practice that are as crucial to quality science and ethics as they are to making music and musical instruments.[25,26] Concomitantly, just as ethics and science share common general roles for diversity and unity within them, so too the diversity between them—for instance, of particular values (empirical accuracy for science, justice for ethics, etc.) and methods (empirical measurement of parameters for science, assessment of autonomy for ethics)—is compatible with their sharing common rational problem-solving processes, which is our contention in this book.[27]

It suffices for our purposes here that such creative construction processes exist, are improvable, and play an essential role in being rational. Creations are rationalized externally by satisfying their constraints deriving from the problem context and by their subsequent interrelations to the other judgment-improving resources for exercising rationality.[28] To conclude with our version of the rational performance standard, it is required of rational science, and respectively of rational ethics, to increase continually, within feasible limits, the variety of normatively useful constructions (e.g., by adding statistical methods to logic, respectively adding institutional ethics committees to decision making), the reach of these constructions across kinds of situations and states of nature (e.g., statistical methods extending from macro to micro dynamics, respectively ethics committee responsibilities extending from family breakdown to corporate reporting and auditing), their sensitivity to informative signals (e.g., to single-neuron activity, respectively to discrimination of stress-resilient from non-resilient family dynamics), and so on. In return, improved construction provides for improved observation and improved methods. This is a core part of the mutually reinforcing upward spiral of normative and informative improvement.

In contrast to formal reason's *narrowing* of the scope of rationality to what is amenable to formal argument, the centrality of creative construction to rational process means that the scope of rationality is in principle *widened* to include all creative design insofar as it contributes to efficacious deliberation. It is not just scientific method, narrowly understood as logical inference that is a candidate for rationality, but the design of good scientific instruments, laboratory protocols, peer review processes, and so on. Similarly, rationality in ethics is not limited to the formal application of ethical rules, but includes the creation of ethics committees, clinical ethicist positions, bioethics courses and programs, as well as institutional design changes and legal responses and remedies, but also well-designed para-legal counseling and mediation, and so on. In short, the upshot of the application of reason in ethics is the design of increasingly sound deliberation, ultimately in support of human flourishing, just as the upshot of the application of reason in science is the design of increasingly sound deliberation, ultimately in support of the flourishing of human knowing. In both science and ethics we thereby achieve an enriched conception of rational intellect and a more satisfying grasp of our problem-solving endeavors.

4.2.4 Systematic Critical Assessment

Discussion of each of the three preceding judgment-improving resources has highlighted the important role of critical assessment, in both science and ethics. For instance, observations cannot be taken for granted, given their limitations and error proneness, but must be supplemented, improved, rechecked for errors, and cross-checked for relevance and accuracy. This can be seen acutely in science in situations like that faced by Galileo where all the normal observations apparently supported the Aristotelian theory and it was what could not be perceived (absolute motion) that was the key insight. It also is acutely visible in ethics, in situations such as that of the women who received genetic counseling (chapter 1), where it was empirical information and value judgments that proved critical because the standard approach to moral decision making was repudiated. Critical assessment shades into improvement; for instance, organized observation, discussed earlier, forms an important part of the critical assessment process, for example, by acquiring missing relevant observations, rechecking previous observations, and arranging for the most suitable perceptual judgment-forming conditions. And, as noted, logic, and methods of reasoning more generally, have an important role in critical assessment; for example, by being used to test for consistency, derive predictions, explanations, and falsifications, reveal assumptions as suppressed premises, and, when mathematically enriched, to construct theories of statistical inference for use in sampling methodologies, theories of error correction, methods for "cleaning" data, and so on. Critical assessment includes carrying out tests, proposing alternative explanations to challenge orthodoxies, checking performance and assumptions in wider ranges of conditions, improving the power of instruments and test methods, arguing the merits of alternative values and alternative assessments, and the like, and applies equally to both science and ethics (with some different orientations and emphases).

Some of these activities can be performed by individuals, but, as we have seen, in most cases individual judgment is powerfully augmented by socially organized processes. The social dimension to critical assessment has a significance that goes beyond simply overcoming individual limitations and imperfections, for it has a role in problem solving that arises out of the conjunction of individual limitation and the power of diversity. Individuals cannot carry out all lines of inquiry (experimental, theoretical, or methodological) or acquire all expertises, pursue all frameworks and values, and so on; indeed, alone they often cannot carry out, acquire, or pursue even one. They must instead, perforce, carve out a sensible personal cognitive trajectory through this welter of alternatives, often with the only kind of coherent purpose available in this situation: to contribute to a superior combined effort.[29] Moreover, there are many situations in which systematic observation and reasoning are not sufficient to reach agreement, because diverse individuals may decide to weigh the available evidence and/or relevant values differently, pursue different potential explanations for unresolved issues, accept different risk tolerances for closing off investigations, and so on.[30] Irresolvable diversity of individual cognitive positions combined with individual fallibility makes critical evaluation necessary on two counts, equally important and interrelated. First, there is the necessity to check individual claims for error. Second, there is the necessity to integrate diverse individual claims so as to construct a coherent public scientific or ethical position.

As already noted, this variety, arising precisely from the variety of roles finite individuals necessarily carve out, plays a crucial role in generating the range of alternative accounts (hypotheses, potential ethical resolutions), investigations (experiments and inquiries), methodological developments, and so on that constitutes the basis for a powerful learning process, scientific or ethical.[31] At the individual level problem solvers demonstrate their rationality by arranging things so that, to the greatest extent possible, individuals follow their own commitments, within institutional constraints, while collectively they exhaust examination of the plausible resolution strategies. Collectively, the remarkable interdependency among individual values (epistemic or ethical) that such arrangements demand requires an institutional explanation focusing on the way in which problem solvers search for, and adapt to, the institutional roles they perceive themselves to play and the way in which these roles are designed in relation to the overall enterprise. Institutionally organized collective goals subsume individual goals but, rather than stifling variety, permit much greater variation among individual supporting roles. In this way science and ethics are inherently social enterprises characterized by fundamental, irresolvable diversity as well as by an extensible web of local and more global agreement.[32]

4.2.5 Synergy of the Four Judgment-Improving Resources

The preceding discussions have anticipated a constructive dividend to the exercise of the four judgment-improving resources: their mutual improvement of each other. For instance, logic and reasoning generally will play an important role in the improvement of observation—for example, for science, by improving reasoning about the theory and practice of optics used in the design of microscopes and other visual aids; for ethics, in using anthropological observation methods to improve observation of relationships and interaction styles; and for both, in analyzing the data on individual perceptual performance and biases in laboratory and clinical settings. In the case of creative construction, tests for logical consistency will, for example, provide important information about, and usually form primary constraints on, new models and theories. And so on. These relationships are entirely mutual. For instance, creative construction contributes to improved logic through the construction of new formalisms, and the finding of counterinstances, often based on observation, can be used to criticize logical claims. Indeed such critical, creative appraisal of logical principles has stood at the heart of the historical process of improving formalizations of logic. It is actually in this way that the use of reason not only improves the content of science and ethics but also our capacity for reason itself. This is the essential bootstrapping synergy of all powerful learning noted above: the exercise of reason both improves judgment and the capacity to improve judgment.

The idea that reason too should thus improve itself must sound strange if reason is thought to be absolute and known a priori, and may still sound strange if it is simply thought to be unitary. But once reason is accepted as a fallible, learnable, internally complex organized procedural skill, with components that may each contribute to improving the others, the idea of self-improvement falls out naturally, and fits well with the social nature of rationality (as discussed later in this section). Of course, reason is also improvable through feedback from the attempt to apply it directly to the improvement of substantive judgment—for instance, as much has been learned about logic from the creation of theories of physics, computation, and expert knowledge as from mathematics, not to mention from the study of reasoning in psychology. The whole system of science and reason forms, at its best, a mutually improving synergistic process of great power. Exactly the same can be said for the relation of reason to ethics, and to social action generally. Indeed, it is the attempt to exercise reason in these areas, frustrated by restriction to logic, that spawned the development of casuistry, narrative, and analogical/metaphorical reasoning (e.g., in law) and other attempts to enrich our methods for improving judgment. To make explicit the underlying naturalism: this conception of an improving rational capacity brings theories and practice of rationality into the same learning dynamic as theories of science and of ethics, in general. The best evidence for deepening objective understanding is the emergence of a synergy of mutual improvement among all elements—skilled judgment and the four judgment-improving resources.[33]

Since marshaling the four judgment-improving resources for rationality requires expertise, and their effective marshaling is typically strongly context-dependent, there is no undifferentiated general individual ability to be rational, in either science or ethics. (There *may* be a general ability to *learn* to be rational.) And since there is wide individual variability in capacity, expertise, and role, plus variability in institutional design and organizational functioning, there is no undifferentiated general institutional ability to be rational either, in science or in ethics. Since expertise, both individual and institutionalized, is learned, neither is there a fixed rational capacity, given, say, at birth; rather, such capacities develop with the individual or institution they characterize. Moreover, the learning process is marked by finitude and fallibility. It is this interactive, open-ended, self-improving process that lies at the heart of being rational.

And, as noted, it is an inherently socially structured process. Rationality requires the socially organized, systematic application of the four judgment-improving resources. Whence our actual capacities to pursue rational science and ethics depend upon judiciously exercising our capacity for apt institutional and cultural design, that is, for design that is cognitively effective and ethically sound. An isolated individual may be able to engage in some forms of utilizing each of the four judgment-improving resources for rationality, but such an individual suffers from genuine epistemic and ethical deficits arising from their finitude and fallibility. Both individual judgment and social organization are essential, and we do not reduce rationality to either alone. Succinctly: individual judgment without social organization is crippled, social organization without individual judgment is blind—but, as preceding discussions revealed, when acting together, mutually reinforcing individual and social capacities form the most powerful feasible frame for rational process.

The social dimension is irresolvably complex for finite, fallible agents. As everywhere within culture, the relation of the individual to the group takes the form of a delicate shaped-and-shaping dynamic: in both science and ethics social process shapes individual development, but individual action also supports and transforms social process.[34] Social goals and process norms subsume individual goals and norms, even while all social normative judgments are grounded in individual normative judgments. As well as whole-institution values and interests, all within-institution subgroups will have their own values and goals to some extent, based in different sub-institutional goals assigned them, competition for external resources and rewards, and so on. This shaped-and-shaping relationship includes many inherent tensions; for instance, the need for both diversity and expertise creates permanent tensions in those who seek to arrive at results in a rational manner. Diversity calls for many partial participants, most inevitably having less expertise, while expertise calls for fewer participants, with the benefit of long learning experiences; both are useful, but finitude insures they are mutually traded off.[35] Resolving the tension is again a matter of judgment because, in advance of a known resolution of the problem at issue, any decision is fraught with risk and/or waste. Sensible choice processes will vary according to the specifics of the situation, and there will be no rules to follow.

The discussions of skilled judgment and the four judgment-improving resources place us in a good position to offer a naturalistic conception of rationality for finite, fallible agents as an antidote to the formalist version.

4.3   Characterizing Rationality

The essential project of reason in the Western tradition is transcendence of ignorance and error of all kinds, achieved through a process of reasoning about empirical experience that improves our state or condition with respect to our ideals and improves our ideals and their surrogates. Thus, for

Jean Piaget, reason just is the capacity to engage in the overall process of organized and organizing improvement, and that is also, broadly, how we understand it.[36] This captures well a natural sense in which reason is central to intelligence.

We begin with these questions: (Q1) what kind of account is to be given of a problem solution or decision being rational? (Q2) what is the source of content for an account of rationality? and (Q3) what should be the relationship of an account of rationality to an account of our empirical cognitive capacities? For the formalist the answers are immediate, respectively: rational solutions follow from logical inference from the initial premises (data, theories, etc.); logical inference is known by transcendental insight; rationality is the normative standard against which empirical knowledge is assessed. However, once we move to a finitist, fallibilist, naturalist approach, the answers are less obvious.

(Q1) *What kind of account of rationality?* Remembering the conclusions reached in chapter 3, we see no alternative but to pursue a process account of rationality: what rationalizes an assertion is its undergoing a particular process and emerging from that process with appropriate credentials. Even formalism offers a quasi process: derivability from accepted knowledge. But an adequate response has to offer a process rich enough to encompass all of the diverse and complex aspects which we have seen are actually involved in problem solving and to bring to bear all feasibly available supporting resources.

(Q2) *What process with what resources?* Following the preceding discussions, we characterize an initial account of rational process as one that provides for all feasible utilization of each of the four judgment-improving resources, along with the tools of compromise and reflective equilibrium formation (chapter 6), to improve judgment in problem solving and decision making. This core characterization suffices for our purposes here. But it is possible to briefly add a little generic elaboration of process.

Following what was learned in chapter 3, the exercise of rationality takes this general form: (I) set up an initial formulation of the problem and of its anticipated solution(s) and the norms against which a satisfactory solution is to be judged; (II) judiciously bring to bear the four judgment-improving resources along with the methods of compromise and wide reflective equilibrium formation as appropriate; (III) formulate a proposed trial solution or partial solution and its attendant decisions; (IV) using (II), develop the proposal to the point where its adequacy can be tested against, for example, its ability, where apt, to generate empirically recognized improvement, its coherence with solutions/decisions made in other, related situations, and so on; (V) return to (II) with proposals that do not deliver sufficient value (perform sufficiently well), to be reworked against discovered sources of deficiency and productivity (failure and success) and applied to (I) and/or (III), and so on through the process until an outcome of sufficient value is found. Finally, add (VI): the quality of the decision-making setup has to be such that it activates the virtues (humility, cooperation, attentiveness, ...) that together constitute what we might call the deep intelligence of the process, meaning that the design of the process exploits as fully as possible all of the psychological constructive resources that humans can bring to bear on problem solving and decision making. (Contrast here the initial and final conditions of the women who received genetic counseling.) Outcomes from a sufficiently rigorously applied process of this kind are considered rational, since there is no more that could feasibly have been done in the circumstances to improve the outcome. (Albeit the outcome and the process that produced it are fallible and open to improvement should that option arise.)

To illustrate, consider brief sketches of how this process applies to science/ ethics/design, in that order. A problem solver (e.g., experimental researcher or hospital bioethicist or architectural designer), seeking to understand what is happening in some situation S, sets up an initial formulation of the problem and of anticipated solutions (what test best reveals S's nature and how can it be realized? or what agreement best satisfies bioethical obligations in S and how are they to be implemented? or what design best satisfies client and public norms for S and how is it to be realized?). Then, judiciously bringing to bear the four judgment-improving resources plus the methods of compromise and wide reflective equilibrium and the deep virtues of investigation, our problem solver formulates a partial proposed solution—for instance, an experimental protocol or an ethically sensitive compromise or an architectural design theme—to investigate and, using these same resources, develops the proposal to the point where its adequacy can be tested against operative epistemic or ethical or client/public norms, including by diverse persons beyond the problem solver. If its performance is satisfactory (if it delivers sufficient value), carry it to completion as the solution. If not, reconsider the proposed problem and solution formulations, the method used and constraints assumed, and the proposed partial solution, using the same resources to identify sources of failure and success. Revisit these steps as often as is required to clearly identify sources of failure and success and thus obtain a sufficiently (epistemically or ethically or pragmatically) valuable outcome.

We devote the next chapter to reconsidering our earlier ethics cases to show the applicability of this approach. We note that, with (II) removed or held tacitly, this is essentially the widely accepted Dorst/Cross/Maher model of design problem solving process.[37] However, the omission of (II) is crucial, as is the absence of higher-order learning in both accounts (see problem-solving model below). Nonetheless, the natural way this process fits science, ethics, and design supports the view that it is the universal core to all rational problem solving. We are not alone in thinking that problem solving might be a useful perspective from which to understand rational process in ethics. There is, for example, the insightful work of Caroline Whitbeck relating online engineering decision making to ethical decision making, to be examined in chapter 6. Kees Dorst and Lambèr Royakkers (2006) also suggest that problem solving in design theory could prove useful to understanding rational ethical decision making—rightly in our view, though they do not go beyond that suggestion to develop any substantive account of rational process the way we do in this book.

(Q3) *What should be the relationship of rational norms and processes to empirical cognitive capacities?* Contrary to the formalist's a priori dichotomy and asymmetry between norms and empirical fact, the normative naturalist holds that both the normative and the empirical accounts are fallible, learnable theories and as such have to be mutually adapted in a respectful two-way iterative adjustment. Adaptation of each theory is driven by success—the empirical theory is driven by success in understanding the actual problem-solving process humans use, both its strengths and its weaknesses, biases, and limitations; and the normative theory by success in guiding the investigation of problem solving, improving our investigatory capacities in the process (cf. Brown 1996). In consequence, the two theories are locked into an iterative developmental process in which improvement in one is used to improve the other. At any given time, our rationality norm—some version of the process I through VI—should reflect our current best cognitive conception of problem solving and consequent decision making, all inherent imperfections noted and compensated or quarantined.

We can provide some promising support for this approach. Based on evidence from problem solving in scientific research and in design, and from the cognitive science of expertise, one of us (CH) has proposed a universal innovative problem-solving model that provides a common core process across all problem-solving domains. This process has the key property that it can solve (if anything can) even the initially ill-defined problems that necessarily lie at the basis of all problem-solving domains, from how to do valid research under fundamentally new conditions, how to design a new kind of object, and how to develop ethics for a new kind of domain, like genetic engineering, to how to develop a new formalism suitable for a class of problems, how to solve a crime, how to throw a good party, or how to become a saint. This core process turns out to be a particular form of the I through VI process described above, thus providing the cognitive counterpart to our rationalizing norm. Its main emendations are (a) to develop an account of changes in problem formulation, method, constraints, solution formulation and partial solution proposals as the main dimensions of change in problem solving and (b) to provide a parallel higher-order process of learning the general character of problem solving in that domain (by remembering, from specific problem solvings, which kinds of problem formulation, method, and so on fail and which succeed). Remembering patterns of success and patterns of failure is central to obtaining higher-order knowledge about problem solving in new problem domains and in turn explains our capacity to develop expertise, including for solving that domain's initially ill-defined problems.[38] We are in a position to adopt this rationalizing process in ethics, as both normative and reflecting our best empirical understanding of the rational process.

4.4 Conclusion

The account of rationality developed here does not provide a finalized definition of rationality, much less any a priori or transcendent sense of normative status. Nonetheless, unlike other accounts that purport to do so, it respects each of the principles and lessons developed in chapter 3. It also unifies the complex cognitive dynamics of scientific and ethical development and provides a naturalist, yet noninstrumental, theory of rationality that is rich and powerful enough to illuminate development of rational capacity itself. It provides a proper basis for understanding the intratwined process of improving rational learning capacity, including self-improvement, while developing solutions to problems—a process central to solving problems. Consequently, its content, while not formally specified, is determined by this scope. Reflexively, this theory of rationality was itself constructed through the same rational process of applying the four judgment-improving resources using skilled judgment, thus it gains its support in the naturalist manner. In ensuing chapters we focus on its application to illuminate ethical decision making and development.

On that score its theme of continuous improvement holds for ethics itself. The general conception of ethics presented inter alia in this chapter is one of the design of flourishing living for all. That conception opens up the space of the applicability of ethics to encompass all dimensions of personal and public life, a conception that dovetails with correlative design processes everywhere, from urban architecture to political institutions. The scope for the improvement of ethics is nothing less than the scope for the improvement of design for flourishing across all these aspects.

5 The Process of Ethical Resolution: Using the Resources of Nonformal Reason

Science and ethics are seamlessly, reflexively theoretical and practical. In the preceding chapters we showed how restricting rationality to formal reason produces distorted, divided, and inadequate accounts of the theoretical and practical dimensions of science and ethics. By avoiding the dichotomies that formal reason creates, nonformal reason integrates theory and practice and thereby provides unified understandings that elucidate and enhance the rationality of science and ethics. In the rest of the book, we develop a broader, more constructive account of ethics that is parallel to the more faithful and realistic account of science that proceeds from nonformal reason. This chapter illustrates how the resources of nonformal reason are used, first by young patients seeking a resolution of a poignantly distressing problem and then as a way of responding to the deep moral challenges posed by several of the vignettes in chapter 2.

5.1 The Rationality and Morality of Dying Children

The ethnographic research of Myra Bluebond-Langner (1978) on how children with leukemia discover, despite a conspiracy of adult silence, that they are dying and then deal with that information in a way that enables them and their parents to retain their dignity, identity, and humanity presents a surprising and illuminating example of the use of the resources of non-formal reason. A troubling moral problem for the parents of terminally ill children and for the health care professionals who care for them is whether to tell them they are dying. Often the children are not told. Why? One obvious answer is the fear that doing so will cause too much worry, anxiety, and distress, and parents want to protect their children as much as they still can. Another obvious answer is that the standard task-oriented rationales for telling the truth to terminally ill adults do not apply. Children do not have to give consent to treatment or to research protocols; they do not have to put their financial affairs in order; and they do not aspire to repair relationships with alienated or estranged family members or friends. Given these differences, the potential benefits of telling might not outweigh the potential harms of telling. But a status-oriented moral reason might be offered in response to such a consequentialist analysis: children are persons, and simply by virtue of being persons, they have a right to know what is happening to them. Even if that claim is correct, the moral agonizing is moot for a practical reason: these children do not have to be told, because they know. That is what Bluebond-Langner discovered: "All of the leukemics that I studied knew their prognosis. All knew that they were dying before death was imminent" (Bluebond-Langner 1978: 165).[1] Two mothers suspected as much, as an exchange between them reveals:

Mrs. Smith: It's pretty hard to keep things from them, no matter how hard you try.

Mrs. Andrews: Have you told your son?

Mrs. Smith: He knows he has a blood disease, and, well, I don't know. But they do pick up quick. (226)

In fact Bluebond-Langner found that children acquired a vast amount of information, not only about the hospital, the people who work in it, and how it is run, but also about their disease, its treatment, its progression, and its prognosis. And she found their acquisition of all that information "even more remarkable because they learned in a situation in which the parents and the staff unconsciously conspired to keep them in painless ignorance" (165). So how do these children "pick up quick"? They are eminently rational, but not in the manner formal reason prescribes or in a manner that formal reason can recognize or explain. Nonformal reason, however, provides an understanding of their remarkable accomplishments.

5.1.1   The Rational Discovery of the Diagnosis

The children whom Bluebond-Langner studied were able to "pick up quick" not just because they are, like most children, curious but because they are, like most children, intelligent and rational. The rationality of these children resides not in what they learned but in how they learned, that is, in the processes they devised and used to discover their diagnoses and prognoses. The resources of nonformal reason figure prominently and essentially in those processes and in doing so illuminate their essential rationality.

Observation   Much of what the children learned about their situations and their surroundings came from careful, intense observation. On their initial entrance to the oncology clinic to which they had been referred for outpatient treatment following diagnosis, they became engrossed in watching children at play in the waiting room. They sat by themselves but had opportunities to observe and question those near them. During their initial clinic visits, Bluebond-Langner reports, "the children, like their parents, watched more than they talked" (176).

In the hospital the children detected a hierarchy of authority among the physicians and were able to construct that hierarchy largely by observing the behavior of the attending physicians responsible for their care and the house staff (the residents and interns) (148–149). They noted other evidence as well. Resident doctors were differentiated because they were younger than the attending hematologists, who were in charge of the leukemic children, but older than the interns, and because, like the hematologists but unlike the interns, they generally wore lab coats over their street clothes. The information the children obtained from their observation not only allowed them to discern similarities and differences, it enabled them to draw helpful inferences. Once they recognized that the hematologists made the decisions about their care, for example, they could challenge a resident who came to perform a procedure by questioning whether the hematologist had authorized it, and after hearing their parents argue with a resident about a procedure, they could stall the procedure by making a fuss about it. The children's acute observation operated more widely. By watching the behavior of residents, interns, and medical students in the evening, for instance, the children were able to identify, with remarkable accuracy, romances between staff members (151–152). And they inferred the likely functions of rooms in the hospital from observations of the equipment in the rooms (136).

These children display an uncanny perspicacity. They know what to watch, how to sift from everything they watch the features of the world that are relevant to them, and, as we will see, how to critically assess their experiences and use them to refine and focus their subsequent observation, i.e., to improve their observing. From the information they obtain from their persistent, dedicated watching, they create classifications and posit generalizations, which they then use reasoning methods to test.

Formal and Informal Reasoning Methods   The children employed familiar reasoning methods to understand their surroundings and what was happening and was likely to happen to them. To begin to comprehend who they were, they quickly differentiated themselves from children with other diseases:

> All the children could distinguish between leukemic and nonleukemic patients. "They [the leukemic patients] come to Monday clinic." "We [leukemic patients]

all have the same blood disease. You know Greta, she comes [to the hospital] all the time. Well, she has a blood disease [sickle cell disease] too, but she goes to another clinic." (155)

Reasoning methods enabled them to explore, develop, and test the significance of their observations. They used induction and analogies to draw inferences from their observations; for example, a boy's loss of hair indicated that he was on the drug vincristine because other boys had lost their hair after being put on vincristine (103). Their reasoning often relied on assessments of other people's behavior: "Alan ... knew that things were not going well when, in spite of how good he felt, his mother would cry. 'In my house, crying and bad things go together'" (180). Similarly, they made comparisons with other people's past behavior to get a sense of the seriousness of their illness:

> The children put people's behavior together with the tests and treatments they were receiving and concluded that they were really sick, "seriously ill," "very, very sick." "This is not like when I had my tonsils out"; or even "when I cut my head open." "This" was somehow worse, as were they. (173)

The children appreciated that arguments by analogy have two sides: the relevant similarities between the items being compared that support the analogy and the relevant dissimilarities between the items being compared that challenge the analogy. When a death occurred or was imminent, the children were concerned with establishing the cause of the death. Having done that, they compared the deceased to themselves. Children who only recently had discovered their prognosis emphasized their dissimilarities to the deceased. But children who had known their prognosis for some time realized that their similarities to the deceased were more compelling (189). In short, they had learned how to improve their informal reasoning.

At times the children used the deduction that typifies formal reason:

> Only children who are dying do not have to go to school anymore.
> All other children must go to school.
> I do not have to go to school anymore.
> Therefore, I am dying. (190)

Deduction, as we will see in the section on systematic critical assessment, is also the logical basis of the hypothesis testing the children did to evaluate the correctness of their inferences.

Creative Construction The children's creative constructions often took typically childish forms. They would congregate in places where they could talk privately, and they would devise strategies for eavesdropping on conversations. Older girls, for example, met in the women's bathroom on the "adult" side of the clinic's waiting room to talk about the disease, drugs, and what was happening to other children. When adults entered, they became silent and often dispersed. Aware of what was going on, boys gathered near the bathroom door (144). Because the children knew that information exchanged between adults was more valuable and more reliable than information adults conveyed to them, they surreptitiously sought that precious information. They would try to delay separations from their parents in the clinic's waiting room so that they had more opportunities to overhear adult conversations; they would hide behind office doors; they would pretend to be asleep; they would turn down the volume on the TV; and they would ask people to be quiet when doctors assembled by the door of their room to talk to their parents. Sometimes they were more daring: one child planted a tape recorder in the doctors' office, which served as a conference room.

But one of Bluebond-Langner's central findings—that the children proceeded through a series of stages of awareness in learning about their diagnosis and prognosis—shows a more sophisticated use of creative construction. She found that the children went through five stages in acquiring information: discovering first that "it" is a serious illness; then learning the names of drugs and side effects; grasping the purposes of treatments and procedures; coming to understand the disease as a series of relapses and remissions (minus death); and finally seeing the disease as a series of relapses and remissions (plus death) (166). At each stage the children had to formulate a new self-concept or definition of themselves: seriously ill; seriously ill and will get better; always ill and will get better; always ill and never will get better; and dying (terminally ill) (169). To move from one stage to the next, the children required both information and experience. When new information was forthcoming, their existing information had to be sufficient to enable them to recognize its relevance and incorporate it into a set of premises from which they could reason to a conclusion:

> Without the requisite information, the children could not integrate new information to come to a new conclusion. For example, if a child knew the names of all the drugs and their side effects, as well as the purposes of various treatments and procedures (stage 3), but did not know the disease was chronic (stage 4), the news of another's death did not lead the child to conclude that the disease had a terminal prognosis. (168)

Experience both yielded new information and linked information:

First, the children needed the disease experiences (e.g., nosebleeds, relapses, bone pain) to gather significant disease-related information. At any sign of illness, the children were taken to the clinic, where they could again meet their peers and discuss what was happening to them. Second, the disease experience enabled them to assimilate this information by relating what they saw and heard to their own experience. (168)

Bluebond-Langner found that children could remain at one stage for an extended time and that "age and intellectual ability were not related to the speed or completeness with which the children passed through the stages" (169). Consequently, some younger children with average intelligence knew more about their prognosis than some very intelligent, significantly older children who were still in their first remission and had made fewer clinic visits. The older children knew only that they had a serious illness. They needed more experience to construct the subsequent stages and realize that they were going to die (169).

Creative construction is a third crucial component of this process because it gives the children the capacity to use their information and their experience to construct and evaluate new conceptions of themselves. They can do this because formal and informal reasoning methods enable them to abstract from the information they have acquired through observation, personal experience, and conversations. Then, systematic critical assessment allows them to test and revise the self-concepts they create until those self-concepts match the information they possess and explain the experiences they have had.

Systematic Critical Assessment   The children conducted systematic critical assessment in two important, familiar ways: to assess the credibility of evidence and to test inferences drawn from that evidence. With respect to evidence, the children astutely realized that information is reliable only if it comes from trustworthy sources. So they tested people who served as their sources, including Bluebond-Langner:

Benjamin asked everyone he saw that day what happened to Maria [who had just died]. Later, when I asked him why he asked everyone, he said, "The ones who tell me are my friends. I knew Maria died. I saw the cart come for her. They told everyone to go in their rooms. I wanted to see if you were really my friend."[2] (188, note 27)

Their critical scrutiny included, not surprisingly, the hospital staff:

The staff's refusal to answer questions directly, and parries such as, "Well, we'll see" or "I don't know," eroded their credibility. "I asked three times when I could go home and they said, 'I don't know.' They saw the tests. They told me that's why [to see when I could go] I had to have them." If the staff did not know, who would? In the children's eyes, the staff had no excuses for not knowing. (180–181)

A hierarchy of trustworthiness emerged for the children. They generally trusted other children because they expected to get straight answers from them; with respect to drugs and medical procedures, though, they excluded children without experience from their discussions, perhaps for reasons of efficiency but more likely because they deemed them unqualified. The only information they prized more highly than information from other children was information obtained by eavesdropping on adults because they knew that adults shared information they withheld from children. They considered information volunteered by adults the least reliable (180).

The children tested the generalizations they inferred from their observations and conversations and the self-concepts they developed at each stage by employing a rudimentary form of scientific hypothesis testing. Bluebond-Langner describes a prevalent practice:

The children tried out their newly acquired self-concept on everyone they saw. Children ... would often exclaim, "I'm really pretty sick, you know," and then offer as evidence the changes in people's behavior towards them or the physical changes in themselves, especially the "wounds" suffered from procedures.... The children would begin by counting their needle pricks and pointing to them. (173)

At stage 1, Bluebond-Langner reports, children used this strategy to begin every encounter, and it continued to be used in the succeeding four stages, albeit less frequently. The repetition at stage 1 probably was important because the children were testing a very different—and undoubtedly frightening—conception of themselves. The outcome of the testing was that "the children found confirmed in people's reaction to their pronouncement the fact that they differed from other children" (174).

The same testing procedure was used to resolve other uncertainties. When a child died or was close to dying, for example, "the children attempted to establish the cause of death either by asking a question or stating a hypothesis and assessing the other party's reaction" (189). In stage 3, after the first relapse made it clear that the disease was chronic, the children were, as Bluebond-Langner puts it, "left more and more to their own devices" (181). They had to obtain the information they wanted from other children, and to do so, "they formulated hypotheses about the relationship between various symptoms and the drugs, procedures, or treatments employed, and checked them out with peers" (181).

The testing procedure the children used is a version of orthodox scientific method: formulate a hypothesis, develop predictions from the hypothesis, and test the predictions to see whether they are confirmed or refuted. It is a drastically simplified and much less reliable version, but nevertheless an instance of a core strategy for generating knowledge in any natural realm.

The children's testing of self-concepts introduces another informal reasoning method: fit or coherence. The progressive growth of knowledge and the unfolding of the stages imposed constraints on how the children could respond to new information and experiences:

> Once children reached stage 4 ... they had fitted all of their acquired information into a pattern of relapses and remissions, of never getting well. And just as all other events could be fitted into the scheme, so too could the death of another child. It could not be seen in isolation, but was part of the same chain of events that they were. They realized that they shared with peers not only the same experiences, but also the same prognosis. (184)

As they acquired more information, the children had to make judgments about how well that new information fit or cohered with their current overall understanding of their situation. When that new information was hard to reconcile with their prevailing understanding or exposed tensions or inconsistencies in that understanding, they had to make further judgments about how to create a new understanding that could accommodate it.

The criterion of fit also operated more specifically. A child's first relapse was profoundly distressing because the preceding months of remission had lured the family into thinking they could lead "normal" lives. With a relapse, the old symptoms were back, and the same medical procedures were being performed. Jarred from their comforting respite, the family had to drastically revise their feelings and their understanding of the situation. In particular, the child had to create a new self-concept that fit the new evidence. People's reactions now were the same as they had been at the time of the initial diagnosis, except now they were crying more and telling less. In addition, the child was no longer getting the drugs for headache and bone pain that had been prescribed during the remission. Instead, the child was put back on the same drugs that had been administered after the diagnosis. To move to stage 3, the child had to create a self-concept that fit these discomfiting changes—and to give up the belief that the drugs had brought about a recovery.

As the children used the resources of nonformal reason, their knowledge improved, and their capacity to use the resources to acquire knowledge improved. With respect to improving knowledge, for example, on their first hospital visit the children immediately began to orient themselves as they stepped off the elevator, quickly making a crude distinction between "us" and "them." "Them" were those in uniform, and "us" were those not in uniform (171). "Us" were the patients and families who took orders and spoke only when spoken to. "Them" came and went as they pleased and inflicted pain or delivered news that caused pain. Initially, the children were not aware of the criteria they used to make these distinctions. They recognized a multiplicity of differences, however, not just the obvious one between doctors and nurses, and they realized that the distinctions were not based solely on uniform. Over time they progressively refined the distinctions to develop an accurate understanding of the hospital staff and their functions.

With respect to their methods, the children were able to critically assess their observational and reasoning skills, adapt them to new circumstances, and hone them as experience and necessity dictated. Bluebond-Langner reports that "as time went on the children became more astute, more capable of drawing conclusions from people's behavior. They knew the indices for various types of behavior and what they meant" (173). The children appreciated how much they relied on observation and testing for information, and as they recognized changes in how people behaved toward them through the progression of their illness, they realized that their observational and testing strategies needed to be adapted and enhanced. In short, experience with using the resources of nonformal reason leads to their application to improving methods of extracting and applying information, making constructions and the like, which is how nonformal reason reflexively improves itself.

This survey of how the children used the resources of nonformal reason highlights four important features of nonformal reason. First, these resources do not function singly and independently; rather, they operate jointly, reciprocally, and reflexively, and in ways that can improve their collective use. Second, the operation of nonformal reason is thoroughly grounded in experience. Watching others and reflecting on one's own experiences are essential to acquiring and testing information and to the creative construction of self-concepts. Third, the operation of nonformal reason is robustly social. Talking, acting out together, and generally interacting and collaborating with others provide the children with crucial information, not just for making judgments about themselves and their peers in various stages of the disease process, but also for testing information and improving their methods of observation, creative construction, and systematic critical assessment. Fourth, the exercise of these skills is also self-improving.

The magnitude of what these children do should not be underestimated. They suddenly find themselves in strange and frightening new circumstances, indeed in circumstances so unprecedentedly different that they have to radically rethink and reorder their environment, their place in that environment, and, ultimately, themselves. They cannot rely on any of their old assumptions or rules, or on any of their old methods; everything has to be relearned. These children manage to do all that. How? Their challenge parallels that of scientists entering novel research environments—for example, quantum mechanical or ape language learning environments—where the basis of understanding is uncertain. In such new environments scientists cannot rely on any of their old assumptions or rules, or on any of their old methods, either; everything has to be relearned. Yet they likewise manage to do so. Again, how? Certainly not through just formal reasoning, which presents only paradoxes of circular assumptions and contradictions from which anything and everything can be deduced. Rather, they use judicious, iterative interaction among observation, creative construction, informal reasoning, and careful critical assessment. The same answer applies (albeit more modestly) to the children.

5.1.2 The Rational Morality of Dying Children

Realizing that children with leukemia know they are dying, and understanding how they know they are dying, might seem to remove any residual moral responsibility to tell them they are dying. Yet it could be argued that the children should be told, both to confirm what they know and to give them an opportunity to talk about death. That position is theoretically attractive, but the lived morality of these dying children defies its abstract appeal. Bluebond-Langner observed the children's "consistent impenetrable silence in the face of ultimate separation and loss" and sought an explanation of this perplexing phenomenon (198). The children are not silent because talking about death is taboo. Moreover, unlike adults who initially conceal a terminal illness but subsequently disclose that they are dying, the children's silence is resolute. Why is secrecy so vital to them?

It is vital because the practice of mutual pretense, which exists when each party knows the patient is dying but acts as if the patient were going to live,[3] preserves the social roles and responsibilities that have acquired magnified importance for the children and their parents, enabling the children to retain their socially created identity as children and to continue to feel a sense of worth. A dying child fundamentally disrupts the social order, because the impending death of a child collapses the established roles of both children and parents (as well as medical professionals). Children are supposed to prepare for, and be prepared for, the future, but a dying child has no future. Clearly the children understood their plight; when they knew they were going to die, they stopped talking about long-range plans and goals:

> They never again mentioned what they were going to be when they grew up, and became angry if anyone else did. A six-year-old boy was angered when his doctor tried to get him to submit to a procedure by explaining it and saying, "I thought you would understand. You once told me you wanted to be a doctor." He screamed back, "I'm not going to be anything," as he threw an empty syringe at her. She said, "OK, OK," but a nurse present asked, "What *are* you going to be?" "A ghost," he replied, and turned over. (194)

Yet practicing mutual pretense is difficult for all parties and sometimes requires quick thinking:

> When a seven-year-old boy's Christmas presents arrived three weeks early and Santa came to visit him, suggesting that the boy would not live to Christmas, he turned to his choked-up family and the staff members around his bed, saying, "Santa has lots of children to see, he just came here first." (204)

What makes the extraordinary vigilance and effort that mutual pretense demands of everybody worth it?

In a society and a culture so thoroughly oriented to the future and so doggedly devoted to achievement, the prospect of death, in addition to the inherent fear and uncertainty it instills, erases a child's identity and concomitant sense of worth. These children appreciate the hopes and expectations their parents harbor for them, and they know they will never fulfill those hopes and expectations. Adults go to work; children go to school. They no longer go to school or even do schoolwork. They will have no future achievements because they have no future. Childhood is about becoming, but there is nothing they will become. Those portentous losses give the children powerful motivation to hold on to as much as they can. Mutual pretense sustains their status as children so they still can be valued as children.

Because the roles of children and parents are symbiotic, mutual pretense likewise sustains the status of their parents so they still can be valued as parents. The loss of authority and control over their children, not just in the hospital but in general, is threatening and debilitating to parents. They feel ignorant: "Many parents thought that they did not know very much, too little to answer their own questions, not to mention the children's" (215). They feel overwhelmed and helpless: "The parents saw themselves as powerless before the disease, the doctors, the machines used for treatment, and, most important, their children. They felt that they could not care for their children on a day-to-day basis, let alone in the future" (215). They lose some of the joys and pleasures of being parents. Mothers could not cuddle their children because being hugged or touched was painful, and most importantly, parents could not nurture, protect, and rear their children, the tasks that define parenthood. By preserving the established social framework, mutual pretense retains order in a world roiled and disordered by disease and thereby affirms the identity and worth of the children and their parents. As Bluebond-Langner emphasizes, "the social order is a moral order" (230).

The children's pretense is indeed deeply caring, loving, and altruistic. Bluebond-Langner comments, "Ironically, these children came to see their own task in life as supporting others" (232). An exchange with one child provides an unsurprisingly childish illustration:

Myra: Jeffrey, why do you always yell at your mother?
Jeffrey: Then she won't miss me when I'm gone. (126)

The children's gift to their parents was, however, also the greatest gift they could give to themselves. The children realized that sustaining the hopes of adults secured their continued presence, which meant that they would not be left alone (228–229). Mutual pretense was a way of satisfying the deep human need to belong:

> Everyone could retain their identity and membership in society, except those who did not practice mutual pretense; they were ostracized and abandoned. To put it another way, as long as individuals fulfill their obligations and responsibilities and do not violate the social order, they are granted continued membership in the society and all that comes with it—freedom from fear and abandonment. No one seemed to know this better than the terminally ill children; for they practiced mutual pretense unto death. (232)

The basic roles, structures, and institutions of society are designed to meet that fundamental need, but at the price of fulfilling the accompanying social responsibilities. In this regard, the loving, caring altruism of the dying children was supplemented and supported by an exquisite appreciation of their vulnerability. For greater than the fear of dying is the fear of dying alone, without comfort, without solace, and without love. The children did not want to be abandoned when they most needed to belong. Their arrival at a moral practice that is practically accessible in the circumstances yet meets the deepest needs of everyone involved is a model of constructive, creative ethical rationality at work; moreover, one that holds its own among the adult achievements of civilized culture and society that support thriving personal lives.

How, then, does the practice of mutual pretense emerge and, given the strenuous demands it imposes on all parties, persist? Conditions propitious to mutual pretense existed from the moment the children entered the hospital. The staff strove to withhold information from patients, to prevent leaks, and to cover up leaks and displayed an unwillingness to answer questions from the children by leaving the room, reprimanding the children, and ignoring their questions (200–201). Parents were reluctant to volunteer information, explained little, and displayed their unwillingness to communicate through lengthy absences from the room, brief interactions with their children, and avoidance of conversations about disease-related topics (201). The children observed this behavior and used informal reasoning techniques such as argument by analogy and induction to recognize the significance of the behavior, to construct patterns, and to infer the implications for their own behavior. Just as they would have countless times before, in their family setting and in the hospital, they recognized and interpreted the nonverbal signals now being sent to them—that adults were not willing to acknowledge their reality and were withdrawing from them—and they responded to those signals by fitting themselves into the changed family and hospital milieux. As Bluebond-Langner notes, "When they realized they were dying and that people were no more willing to talk about their condition than they had ever been, the children started to practice mutual pretense" (210). Although engaging in mutual pretense is a different kind of children's pretendings, it was readily accessible to them because it was a consoling adaptation of their experiences with their parents and the hospital staff.

If the process by which children arrive at pretense explains only its general character as an intelligent adaptation, the process of continuing pretense reveals more. Pretense is not a simple single action (deny I'm dying) that is repeated. Rather, pretense is a whole complex, contextual, and reciprocal pattern of relating that is needed to express denial and that must be continually assessed and reconstructed as circumstances change. How the children managed to maintain that pattern again displays their use of nonformal reason. The practice of mutual pretense involves following guidelines such as avoiding dangerous topics, keeping interactions brief, and ignoring emotional breakdowns, and the children devised tactics accordingly. Children who had never cried about painful procedures began to refuse the procedures by crying, in the way "normal" children do, instead of protesting, "Well, I'm dying anyway," or "It won't do any good" (204).

A boy and a staff member talked about past Christmas plans as if they were for an upcoming Christmas even though he was unlikely to celebrate it (202). As did their parents, children adopted a "distancing" stratagem to avoid or limit interactions with others. But because they were confined to their beds, the children could not avoid being in a room or make excuses to leave, so they had to refuse to talk, pretend to be asleep, lash out or cry out continually, or speak inaudibly or superficially (206–207). Whereas the children had formerly employed tactics for cultivating relationships, they now adopted tactics that would weaken and destroy relationships, to reduce the number of people with whom they had arduously to sustain the illusion that they were not dying.

Preserving mutual pretense is, as Bluebond-Langner describes it, a "delicately balanced drama" (208). Preserving that precarious balance requires constant attentiveness and perceptiveness and the ability to make specific contextual judgments about competing demands. The children must be disciplined yet accommodating; committed yet flexible; strong yet sensitive; self-interested yet altruistic. Making good judgments involves the resources of nonformal reason: scrupulous observation of their surroundings and of others' demeanor and reactions as well as their own demeanor and reactions; sound inferences about what has happened in the past and why it happened, what is likely to happen now, and what would be effective means to desired ends; creative responses to breakdowns and ways of preventing breakdowns; and unrelenting, comprehensive critical assessment of particular aspects of the practice (e.g., why did I slip up there, and how can I avoid doing that again?) and the practice as a whole (e.g., are people going along with me, or are they questioning and challenging me more?). The success of the children—that, as Bluebond-Langner reports, "they practiced mutual pretense unto death" (232)—testifies to their perspicacity. The children certainly were "willful" and "purposeful," and their use of nonformal reason enabled them not only to act in the world created for them but to create their own world, the world that they and their parents needed.

The justification of mutual pretense does not come from the application of a philosophical principle or theory. Mutual pretense is a complex, morally rich, socially embedded, life-affirming practice. No formal analysis could have deduced from unnuanced, blunt universal rules a position that subtly extracts moral value from ultimate loss through the subterfuge of a tacitly agreed lie. This is an unexpected, powerful—and wincingly beautiful—response that does moral justice to all parties. Only the whole intelligence, using all its informal resources, can arrive at so profound a resolution. Nonformal reason is the capacity that makes such outcomes humanly possible.

So, is practicing mutual pretense the right way to respond to the impending death of a child? That is a formal-reason question, particularly when asked without regard to the experiences of those involved in the practice or the diverse contexts within which the practice operates. Can the practice of mutual pretense be improved? Is there a better way to respond to the impending death of a child, perhaps a hybrid that preserves the identity-forming, worth-creating roles of dying children and their parents yet allows the children to talk about dying? Those are the questions of nonformal reason, which is always critically attuned and searching for improvement. If there is a better way, it will not come from an abstract moral principle or moral theory. It will come from experience honed by careful observation and systematic critical assessment of and reasoning about mutual pretense, complemented with creative construction of novel alternatives. And ascertaining its superiority will be a matter of judgment. Above all, though, it will respect what the real rationality and morality of these children teach us.[4]

The dying children with whom Bluebond-Langner worked were richly rational and heroically moral. They were understandably eager and determined to find out what was wrong with them, and they used the resources of nonformal reason to acquire the knowledge they sought. They had an impressively mature and sophisticated appreciation of the impact of their illness on their parents, and they used the resources of nonformal reason to find a way of preserving the identity, dignity, and worth of their parents and themselves. They displayed seemingly preternatural moral acumen, sensitivity, and fortitude in maintaining, in the direst of personal circumstances, the silence of protection. They knew that the process of dying is for the living, and they knew how to guard the living.[5]

5.2 Using the Resources of Nonformal Reason in the Vignettes

Chapter 1 presented women's process of deciding whether to try to become pregnant after receiving genetic counseling as a paradigm of skilled deliberative judgment. In this section we draw upon selected vignettes from chapter 2 to illustrate how nonformal reason operates in various other processes of rational moral deliberation.

5.2.1 Ms. B

The vignette about Ms. B (see vignette 2.1) was chosen to show how the meaning of crucial moral terms—in this case "euthanasia"—can be multiply ambiguous, and that they cannot be intelligently applied until they are contextually disambiguated. The resources of formal reason are too meager for this task. How does nonformal reason manage it?

Ms. B is afraid of dying by suffocation, so she has chosen terminal sedation as the means by which she will die. This choice seems clear to Ms. B and consistent with her Christian values until a family friend interprets the arrangement as euthanasia, creating confusion and dismay about whether her plan violates the religious and moral convictions of Ms. B and her family. The method of applied bioethics offers little help to Ms. B and her family, and little direction to Dr. M, because the concept of euthanasia is vague and indeterminate. Timothy Quill, Bernard Lo, and Dan Brock (1997) demonstrate that, in practice, various forms of ending life are hard to distinguish from one another, so determining which ones to categorize as instances of euthanasia is uncertain and controversial. In response to that difficulty, Quill, Lo, and Brock shift from a moral principle about euthanasia to a moral principle about respect for autonomy. For them, the core moral issue in end-of-life decision making with competent patients is whether autonomy is respected, that is, whether the consent of the competent patient is voluntary and informed. But their alternative elides the real problem because Ms. B's difficulty lies not in consenting—an expression of autonomy—but in the moral reflection and effective deliberation that give substance and meaning to personal authenticity, a fully realized sense of autonomy.

To be sure, informed, voluntary consent has become the most important facet of autonomy.[6] "'Autonomy' thus means freedom from external constraint and the presence of critical mental capacities such as understanding, intending, and voluntary decision-making capacity" (Beauchamp 2003: 24). But this is a narrow, thin construal of autonomy. Richer is Bruce Miller's (1981) four senses of autonomy, discussed in chapter 2. Informed, voluntary consent is incorporated into two of the four senses—autonomy as free action and autonomy as effective deliberation. Requiring that a patient give permission for a proffered treatment in the absence of undue influence, manipulation, or coercion makes the patient's decision voluntary; having information about a proffered treatment makes the patient's decision intentional; and both together make the patient's decision free. Awareness of the alternatives and the likely consequences of those alternatives allows a patient to evaluate options and come to a decision based on that evaluation, which is engaging in effective deliberation. Effective deliberation in this sense will be recognized as a weak form of the wider, more powerful deliberation process used by the women after receiving genetic counseling and fully supported by nonformal reason.

A crucial issue for autonomy construed as effective deliberation, however, is how effectiveness is determined. One way in which deliberation can fail to be effective is that the patient's weighting of values can be nonrational, which, Miller explains, means "either that the weighting is inconsistent with other values that the patient holds or that there is good evidence that the patient will not persist in the weighting" (1981: 25). Although Ms. B's problem is about values, it is not yet a problem of inconsistency between her values or of vacillation in adhering to her values. Until she was forced to, Ms. B probably never had envisaged that she would have to decide about the management of her own death. As well, Ms. B probably had never thought about euthanasia in more than vague, superficial terms, and when she did, she likely had never entertained the possibility that she would request euthanasia. When she had to decide how she would die, she assumed that not dying by suffocation is an intelligible, desirable, and morally permissible end and that terminal sedation is a morally innocuous means to that end. She believed that aggressive sedation would simply render her unconscious and thereby prevent her from experiencing the horrors of dying by suffocation; she did not appreciate that aggressive sedation also could depress her respiration and thus could cause her to die sooner than she would without it.

Ms. B's assumptions have been thoroughly roiled. Her goal of not dying by suffocation now is potentially constrained, if not precluded, by a norm that prohibits euthanasia. Because her situation has become complicated, confusing, and uncertain, she needs a thorough, comprehensive, critical examination of her plight. In that regard, Ms. B's problem engages the two deeper, more complex senses of autonomy that Miller distinguishes: autonomy as authenticity and autonomy as moral reflection.

Effective deliberation can occur without any questioning or assessment of the values utilized in the deliberation, but not for Ms. B. As part of her deliberation, Ms. B must formulate the values that have been emblematic of how she has lived and are important to how she continues to live and how she will die. Given her unanticipated and novel situation, this is more than an exercise in values clarification; rather, it is an exercise in the construction and evaluation of values as an integral part of the construction of a self. Ms. B will be testing and critically examining the values she entertains to see how well they fit her past life and her new circumstances. She will be using the resources of creative construction and comprehensive critical assessment, and those resources will reciprocally enhance one another. As Miller recognizes, this sort of deliberation unites the important but neglected senses of autonomy as *authenticity* and autonomy as *moral reflection*: "Moral reflection can be related to authenticity by regarding the former as determining what sort of person one will be and in comparison to which one's actions can be judged as authentic or inauthentic" (1981: 25).

What Ms. B needs for her imposing task is not just moral insulation—independence and protection from interference in her decision making—but moral help. She needs help understanding why she unexpectedly has a problem, what her problem is, and how she can work through it. Helping her is, in part, a matter of conveying information, but it will not be enough to simply give her information and tell her to choose what she wants. Franz Ingelfinger, a longtime editor of the *New England Journal of Medicine*, once said, with vigor, "A physician who merely spreads an array of vendibles in front of the patient and then says, 'Go ahead and choose, it's your life,' is guilty of shirking his duty, if not of malpractice" (1980: 1509). Ingelfinger stresses that a physician must take the responsibility of recommending a concrete, practicable but also morally insightful and satisfying course of action to a patient; but no such recommendation can be made to Ms. B without an understanding of the kind of person she is and wants to be in her death.

The goal for all those helping Ms. B, but particularly for Dr. M, is to avoid a looming tragic outcome. Do moral and religious prohibitions of euthanasia prevent her from dying in the manner she desires? What, exactly, are the nature, the scope, and the force of those constraints? As the women who after genetic counseling are searching for a green light to try to conceive, so is Ms. B searching for a green light to avoid dying by suffocation. Dr. M's responsibility, because he is the physician caring for her and because he is the one who ultimately will have to endorse or reject her decision, is to help her, and at times perhaps guide her, through her search.

Both Ms. B and Dr. M will be active participants in that search and will be engaged in an on-going, reciprocal process of making judgments that are informed, shaped, and tested by the resources of nonformal reason. Dr. M will begin by making judgments about substance and process: about what is troubling Ms. B and, given that, about what is the best way to try to help her. Ms. B also will begin by making judgments about substance and process: about the relevance, plausibility, and importance of what Dr. M tells her; about the significance of the way she is reacting to what she learns from Dr. M; and about how she should go about making her decision.

Their collaborative search requires a comprehensive critical assessment of the elements of Ms. B's dilemma. That scrutiny should begin with a strong reaffirmation of the mission of palliative care: to identify and mitigate suffering of whatever kind. Palliative care is about facilitating a good death, and systematic critical assessment can emphasize that that is what *euthanasia* literally means. In that general sense it is hard to imagine that anyone would not accept euthanasia. So Ms. B must fear that there is some specific subordinate issue that clashes with her religious values. That fear has to be investigated. Dr. M then can explain how means and ends are inseparably entwined in decisions about how someone will die and how the interdependence of means and ends can create confusion with respect to the vague and emotionally charged notion of euthanasia; and he can assure her that he will support her throughout her search for personal clarity on the issue and respect her eventual wishes.

In that process, for instance, the notion of "playing God" will undoubtedly be subjected to systematic critical assessment. What is the proscription of "playing God" supposed to prohibit? A death that is not "natural"? If so, what does "natural" mean? Or just dying sooner than one would have without intervention? If so, what kind of intervention? What if both providing privacy for Ms. B in a single room and providing her with company in a multi-occupant room were to lead (from, respectively, loneliness and shame) to depression which hastened death; would either be an inappropriate intervention? Or does "playing God" mean exercising a kind of control over life that one should not exercise? If so, is the choice of a room different from the choice of sedation? Or perhaps "playing God" is having an intention or a desire that one should not have? Again, is the choice of a room different from the choice of sedation, and if so, why? And similarly for all the other possible kinds of intervention, control, or intention, why would they be supposed to be wrong?

Ms. B's ideals and the values that could serve as proxies for her ideals also must be explored critically. What does she want? Suppose Ms. B says she wants to be a moral person and a devout Christian and she wants a good death. Those are compelling ideals, but rather empty when taken in isolation. What specifically do they mean to Ms. B? A good death probably means dying peacefully, without suffering and with dignity, but what being a moral person and a devout Christian means is less clear. If the latter means not "playing God," for example, we can appreciate the complexities in constructively specifying what Ms. B seeks. Values that serve as proxies for her ideals must be identified, and then developed and refined by determining how the alternative ways in which she could die do and do not instantiate them. Ms. B will make judgments about which values are more compelling to her, whether her values are mutually compatible, and where her values are leading her; about how well she is enduring the responsibility of having to decide; and about whether her deliberation is proceeding in a helpful, productive manner. At the same time, her clinical options can be modified and adjusted in the effort to try to find a way of dying that comports with her crystallizing values. Perhaps a little creative construction might help her here: if an explicit protocol were developed that initially permitted her to set her own sedation level—to a level where she felt she was sufficiently comfortable—without the intention to relieve her pain by foreshortening her life, which would be against God's will, and with the assurance that a level of sedation that would keep her sufficiently comfortable would be continued when she could no longer do so, might this arrangement be acceptable to Ms. B? It would, for example, satisfy the Catholic doctrine of double effect. There will be a complex, dynamic interplay between her values and her options, a process of reciprocal testing and customization aimed at designing a way of dying that satisfies Ms. B's values and thereby preserves her ideals. The resources of comprehensive critical assessment and creative construction will direct this crucial component of the process.

Throughout the entire process Dr. M will be observing Ms. B and making judgments about how she is responding, intellectually and emotionally, to information and to their interactions; how she is assimilating and assessing material and reasoning with it; and how she is coping with the burden of having to make a decision. Does she retain information? Is she thinking clearly? Are her emotions appropriate? Based on what he observes, he might need to revise his views about what is troubling her, what kind of further help she needs, and who best can provide that help. His observations and the inferences he draws from them should be compared with and checked against the observations and inferences of Ms. B's family and the other health professionals who care for her. Based upon their observations of how she looks, how alert she is, what her moods are like, and how she is sleeping, for example, what is their sense of how well Ms. B understands her situation and is working through it, how she is feeling, how confident she is about what she wants, and what kind of help and support she needs?

At the same time, Ms. B will be observing the behavior and demeanor of Dr. M. Is his communication of information candid and forthright, or does it seem hedged and evasive? Does he readily acknowledge weaknesses and uncertainties? How does he respond to requests for clarification and explanation and to being questioned about his judgments? Is he receptive and supportive, or does he become defensive? Ms. B will check the information he gives and the opinions he offers against what he has told her family and against the views of other health care professionals. In light of her observations and the evidence she obtains from her interactions with others, she will continually be making judgments about Dr. M's forthrightness and trustworthiness and the soundness of his opinions. Ms. B also will be observing the behavior and demeanor of her family. How are they coping with her plight? Do they look anxious and worried? Are they uncharacteristically silent or nervous around her? Are they edgy and short-tempered? Or, on the other side, are they infuriatingly passive and deferential? Ms. B will try to gauge how much of a toll her predicament is taking on them, and judgments about that, too, will figure in her deliberation.

Formal and nonformal reasoning methods will pervade the process. Dr. M can question how and why Ms. B arrives at her views and whether the conclusions she draws from them really follow, and he can probe the consistency of her positions. He can use narrative and case-based reasoning and argument by analogy to examine the various methods of ending life that are difficult to distinguish and to relate them to Ms. B's values. Dr. M can recount stories about patients in similar circumstances that reveal how and why they made their choices and what happened to them. He can describe scenarios that differentiate interventions used to instigate the dying process—for example, for patients who have unremitting, unbearable, all-consuming pain or patients for whom life has lost all meaning—from interventions used not to accelerate death but to change the nature of the dying process.

Creative construction functions substantively in the operation of the process and procedurally in the design of the process. Dr. M has to be flexible and imaginative in responding to difficulties Ms. B might be having, reframing matters for her, devising new scenarios for her to consider, and altering old scenarios to illustrate relevant similarities and dissimilarities to her particular circumstances, and in entertaining novel clinical options for her (such as the sedative delivery protocol suggested above). Ms. B has to be flexible and imaginative in filling in her ideals of a moral person, a devout Christian, and a good death, and in engaging the values that come to serve as proxies for these ideals with the cases, stories, and possibilities that Dr. M provides.

All the substantive work occurs within a flexible, adjustable process that Dr. M devises and Ms. B accepts, and that both Dr. M and Ms. B continually appraise and modify to improve it. For their work to be productive and helpful, the process must be collective and consultative. Dr. M should involve Ms. B's family so that their fears too can be addressed and so that they can provide sympathetic support for, and a trustworthy check on, Ms. B's deliberation and judgment. Dr. M should also enlist the help of a religious advisor who can guide Ms. B in developing theologically sound and personally meaningful proxies for the ideal of being a devout Christian and in relating her clinical options to the doctrines and values of her faith. A clinical ethicist not hamstrung by the model of applied ethics can do the same on the secular side. And Dr. M should involve experienced palliative care physicians in his exploration of clinical alternatives that manifest, as much as possible, the values of Ms. B. This collaborative, critically attuned process will continue until Ms. B and Dr. M are able to make judgments about when and how it can end—until Dr. M is able to make a recommendation and Ms. B is able to make a decision.

Ms. B's deliberation is much more complex than is recognized by the conventional account of respecting patient autonomy (and the accompanying legal account of obtaining informed consent). It is not, as that orthodoxy implies, simply an individual matter; it is a social endeavor that can be rationally designed to enhance the rationality of the judgments Ms. B makes and the decision that emanates from those judgments. The process is collective for two reasons. One is that, as we have seen in science, the resources of nonformal reason are more effective and more powerful when they operate socially. With respect to observation, the more different modes of access to how Ms. B is thinking and feeling there are, and the more the outcomes of that multiple access can be comparatively assessed, the better can be the judgments of others about how she is doing and what she needs. Likewise with respect to critical assessment and methods of reasoning. The more diverse resources and perspectives that can be brought to bear, in a coordinated, focused manner, on understanding her problem and ways of dealing with it, the more confident one can be about the rationality of the process and thus the outcomes of that process.

The other reason is that, contrary to what the conventional understanding of autonomy suggests, Ms. B's values are not individual and private. Ms. B did not create her ideals and values de novo. Her ideals and values have emerged within multiple historical and social contexts, most notably and influentially those of her family and her religion, and have been tested in countless social interactions and adapted to a multitude of social circumstances throughout her life. Because values are socially created and grounded, stories about and illustrations of their genesis—nonformal reasoning techniques—are central to understanding and assessing their legitimacy. In addition, given the social genesis, meaning, and operation of values, it is eminently reasonable for Ms. B, as did the women deciding whether to try to conceive after genetic counseling, to search for the social boundaries within which she may decide. That search should not be misconstrued as an abandonment of her autonomy; rather, constructing a realistic strategy for framing and managing her problem should be embraced as an expression of her autonomy. The stories and perspectives that Dr. M and others provide can enable Ms. B to formulate and assess her values and to appreciate the limits within which her values will operate.

The point of this posited depiction of Ms. B's deliberation and the insistence on its inherently social nature is not to denigrate or disparage autonomy—autonomy is an important moral ideal. The point is to comprehend the complexity of autonomy. Like the women deciding whether to try to conceive after receiving genetic counseling, Ms. B's predicament is profoundly and momentously individual. And like the women deciding whether to try to conceive after receiving genetic counseling, Ms. B's deliberation is robustly social. The challenge is not just to respect her right to decide but to help her make the right decision—the decision that is right for her—by deciding rightly. Deciding rightly is deciding rationally.

The entire process that Ms. B and Dr. M develop is a creative construction designed to resolve an initially ill-defined problem in a manner appropriate to the issues and the persons involved. This accomplishment is at least one of the most important exercises of rational intelligence in the moral domain, if not the most important. It necessarily underlies all that transpires at more practically accessible levels of moral discussion and action. Formal reason lacks that capacity. Nonformal reason has that capacity and also illuminates and rationalizes it.

Ms. B's struggle might not end well. Rationality, as Bertrand Russell reminded us, is about thinking and acting intelligently; it does not, and cannot, guarantee nice outcomes (1912/2005, chapter 6). At the end of her deliberation, Ms. B's situationally tempered values could be incompatible (as in the example of selecting a room for Ms. B above), and she could face a no-win choice. In addition to her fear of suffocation, she now fears that using the means necessary to avoid suffocation would violate her deep moral and religious convictions, which represent what it means for her to be a moral person and a devout Christian. After careful scrutiny and assessment of her situation, one of those fears might persist. If so, she will have to choose between not dying by suffocation but doing something, just as she is dying, that she regards as fundamentally wrong, and abiding by the precepts of her morality and her faith but enduring the kind of death she fears so profoundly. In that event, no matter what Ms. B chooses, she will lose something vital to her conception of what it means to be a good person and to lead a good life. But a process of deliberation shaped, tested, and conducted by the resources of nonformal reason gives her both an examined death, as an expression of her self, and a better prospect of a life that ends as well as it can.

5.2.2   Ms. F

The preceding vignette about Ms. B shows how the meaning of a core moral term—*autonomy*—is ambiguous and cannot be meaningfully applied until it is contextually disambiguated. In vignette 2.2 Ms. F has been offered the option of a third lung transplant, and she has to use the resources of non-formal reason to make an autonomous choice in circumstances riven with uncertainty. Refusing the transplant means death. Accepting the transplant offers the possibility of continued life, but the probability of a successful transplant is uncertain, if the transplant is successful, how much longer she might live is uncertain, and what that longer life would be like is uncertain. In general, what would she have to sacrifice and endure to have a life of unknown quality for an unknown time? Overlying those issues are the personal and social considerations. Pressures from her family and those close to her push in both directions, so no matter what she decides, she will disappoint and dismay people who are important to her. Moreover, if she decides to try another transplant, she has to move to a different hospital and lose the transplant team that has become a "family" to her. Yet how can she refuse? How can a person facing imminent death choose to die when given an alternative, even if the procedure is risky and the benefit is speculative? The clinical ethicist who worked with Ms. F was appalled that such a choice had been imposed on her: "I was outraged with the physician who had made what I thought was an outlandish and exploitive, if not coercive, offer of a third lung transplant" (Marshall 2001: 147).

How might the clinical ethicist help Ms. F? Her predicament is not amenable to the decision making of applied ethics because it is not a matter of applying or adapting a consistent set of principles to a problem, nor is it a matter of maximizing utility on the basis of specified risks and values, none of which are even well defined here. Nor is it just a matter of determining whether Ms. F's decision, whatever it turns out to be, is autonomous in the spare sense of being free and informed. What the clinical ethicist can do is encourage her to assess critically how she is framing her choice and her process for making a choice, and that examination could be instigated by focusing on her analogy between a broken-down bicycle and her broken-down body. Why does that mechanical conception of her body appeal to her? Is a person simply an assemblage of material parts and a transplant surgeon just a body mechanic? Does she regard that metaphor as a realistic way of thinking about her dilemma, or does it represent and confirm what she already knows she has to do? It always is hard to refuse a gift, perhaps unthinkably so when it is the gift of life. If so, Ms. F is entrapped. But if she is unsure, perhaps all she can do is defer agreeing to the transplant and hope that something will change in a way that frees her, either by clarifying the decision or by preempting it. Making that decision, and accepting responsibility for that decision, could be too hard. Comprehensive critical assessment might dislodge her resignation—the perception that she has no choice. If so, the clinical ethicist could suggest that Ms. F imagine different scenarios of what her life might be like after a transplant and assess her reactions to those scenarios, including their effects on her relationships with the people who are close and important to her, a project for which she has abundant experience.

For that enterprise to be productive, though, the conflicting personal and social pressures imposed on Ms. F have to be removed or at least eased. Doing that requires designing a process that enables the two sides to understand that they are exacerbating Ms. F's dilemma and encourages them instead to explore ways in which they can support her through her ordeal. The clinical ethicist could enlist a social worker or a member of Ms. F's health care team to facilitate separate meetings with Ms. F and members of each side to explore the complexity and the difficulty of her situation and the role they play in it.

Ms. F's decision is daunting because she does not want the control and the responsibility the decision gives her. Requesting to die when one is suffering or life has become meaningless is hard, but refusing a chance to live is even harder. Perhaps Ms. F needs permission to refuse the offer of a third transplant. Nobody, including the clinical ethicist, can give her that permission. What the clinical ethicist might be able to do, though, is design and guide a process of rational deliberation that could allow Ms. F to give herself the permission she is seeking. However it all turns out for Ms. F, a deeper and wider process of deliberation, like the one used by Ms. B, needs to be constructed for Ms. F. Only nonformal reason's rich sense of design for flourishing—even in the most straitened circumstances—can create such a process that fits Ms. F's capacities and her situation.

The broader moral issue that Ms. F's struggle raises is prevention. Wherever possible, moral problems should be prevented, not just solved. What aspects of the institutional and structural design of a health care system made the offer of a third lung transplant to Ms. F possible? Speculating about mercenary or self-aggrandizing motives on the part of the transplant surgeon would be distracting and would evade the deeper institutional design issues that underlie Ms. F's problem. What institutional roles and relationships allowed the departing physician to act unilaterally? Why did the offer not have to be vetted and approved by the team that had cared for Ms. F for so long? What professional and economic incentives drive the establishment of new transplant programs and spur recruitment of patients for those programs? And what social and cultural ethos underpins the health care system and its constitutive institutions and practices? As Ms. F's clinical ethicist observes, "This situation could have arisen only in a clinical setting in which the biomedical resources, technology, and professional capability support the ideology—to a certain extent, even the mythology—that human bodies can live indefinitely" (Marshall 2001: 147).

The individualistic orientation of applied bioethics neglects background ethical issues and, in any event, does not have the resources to address them. As we will see in part II of the book, nonformal reason promotes a more expansive conception of morality that takes ethics to be fundamentally a matter of design, of public as well as private matters.

5.2.3   K'aila

The story of K'aila (vignette 2.3) vividly demonstrates how moral depth or insight transcends the conflicts among moral principles. It is unclear what course of action is in K'aila's best interest, because there are medical rationales both for trying a transplant that could extend K'aila's life, albeit only modestly while also risking losing it, and for providing compassionate care to K'aila as he dies. K'aila's parents must make a judgment not only about what is best for K'aila but about what course of action they can live with, that is, how they can preserve the integrity of the persons they are. Momentous decisions in life—for example, choosing a career, getting married, and having children—are ultimately decisions about the kind of person one wants to be. Considering their integrity does not mean, however, that the decision of K'aila's parents will be egoistic or selfish; to the contrary, their concern for the welfare of K'aila is even more paramount because K'aila is an intrinsic and important part not just of their lives but of who they are.

Individually and socially organized processes for the regulation of moral judgment enable K'aila's parents to deliberate rationally throughout their struggle. The individually organized processes take several forms. The reasoning that K'aila's mother describes occurs at multiple levels, for instance, substantively at the levels of K'aila, themselves, and society and procedurally at the levels of deliberating and deliberating about how they are deliberating. Those levels operate reciprocally and reveal the nature of their personal integrity to be the integrity of integration and coherence, as K'aila's mother recognizes when she comments, I "could not deny what I knew at other levels." K'aila's mother reports that her reasoning was checked by her feelings; for example, she "felt lucid," an observation that shows that the kind of integrity being sought is the integrity of an entire human being. And her butterfly metaphor is both the culmination of her reflection and reasoning and an affirmation that she has arrived at a decision with which she can live and for which she can take responsibility.

Several socially organized processes provide checks on the individual reasoning of K'aila's parents, and these processes operate within various institutional mechanisms. One check comes from the pediatrician who cared for K'aila, a responsibility that permitted him to form his own view about what was in K'aila's best interest and to pursue his view when K'aila's parents chose differently. Another comes from the child welfare agency that became involved at the instigation of the pediatrician and that performed its role as an agent of the state in protecting children. Then there are the requirements and procedures of courts in resolving disputes between individuals and the state about the welfare of children, and the individual judges who hear evidence about, reason through, and resolve such disputes. In their deliberations K'aila's parents had to take account of and respond to the demands and constraints of these socially organized processes for regulating individual judgment.

The resources of nonformal reason are utilized in the conscientious efforts of K'aila's parents to decide what to do for K'aila. They mostly use informal reasoning methods, which are subsequently integrated with formal reasoning methods when the judge invokes social and legal norms to arrive at a legal ruling. Creative construction is exemplified in the butterfly metaphor as a way of focusing the moral issue for them as they search for moral coherence and integrity in their lives in the face of weak and inadequate social guidance. Their search for coherence across the types and levels of their reasoning, singly and jointly, displays comprehensive critical assessment. In addition, the activities of the pediatrician, the child welfare agency, and the judge provide external social checks on their deliberation. The upshot is a rational process of individual decision making that is embedded within and constrained by a rational, institutionalized process of social decision making.

5.2.4   Mrs. Smith

When, for many reasons, the ethical considerations relevant to a situation do not provide adequate guidance about what ought to be done, in practice the focus shifts from making the right decision to making a decision in the right way. The substantive question of what ought to be done is replaced by the procedural questions of who ought to decide and how they should go about deciding. That is what happened in K'aila's case, and that also happens in Mrs. Smith's predicament (vignette 2.4), but in an instructively different way.

The challenge of deciding whether to insert a feeding tube into Mrs. Smith's stomach displays the indeterminacy of the principles formal reason brings to this issue: the substituted judgment principle and the best interest principle. Mrs. Smith's situation has a number of similarities to K'aila's. In each, a life-and-death decision is being made for someone who is incapable of such a decision. In both, the problem is resolved by shifting from substance to process. And in both, the potential for supervision and intervention by health care professionals and courts imposes robust socially organized constraints on individual decision making.

Yet there is an important dissimilarity between the two situations. With respect to Mrs. Smith the core values that would make the decision clearly moral are hard to discern. Mrs. Smith is not suffering. Mrs. Smith can no longer interact with other people or the world. She has lost the capacities that are constitutive not just of autonomy but of agency itself. There is no identifiable family ethos, tradition, or set of values, fidelity to which would provide a way of formulating and understanding the importance of what is at stake and would give direction for making a decision. Nor does vitalism—the concern to preserve life because it is life—seem to be operating here.

In the probable absence of such core values, the considerations that remain appear to concern pragmatic matters and the interests of family members. To be sure, just as K'aila is an intrinsic, important part of the lives of his parents, Mrs. Smith is an intrinsic, important part of the lives of her children. But whereas for K'aila there are profound questions about the means and the prospects of keeping him alive and even more profound questions about what it would be like for him to be alive, for Mrs. Smith there are neither. And whereas for K'aila's parents there are profound questions about how their decision can be made in a way that preserves the coherence and integrity of their own lives, for Mrs. Smith's children there is no evidence of a comparable challenge to them individually or to their filial harmony. Perhaps, then, Mrs. Smith's predicament does not raise a moral problem. And perhaps that is why it falls in what Arras (section 2.3) calls the gray area.

Given those telling differences, an individual moral decision about whether to place a feeding tube for Mrs. Smith might not be amenable to either formal or nonformal reason. What is available only to the latter, however, is making the judgment that there is no moral issue at stake. The only further way nonformal reason might operate here is through the socially organized regulation of judgment, provided by health care professionals and a court, in the event that the family's decision making transgresses established limits of social and legal acceptability.

Nonformal reason can fail to provide adequate guidance for moral problems, but that happens less quickly and less frequently than with formal reason because the meager resources of formal reason have two endemic, debilitating problems: the vagueness and ambiguity of its principles and conflicts among its principles. Either there are no higher-order principles from which solutions to these two problems can be deduced, or, to cover all the conflicts, there must be multiple higher-order principles that, a few moments reflection reveals, will also have conflicts and, in turn, need higher-higher-order principles, and ad infinitum. Similarly, the shift from substance to process will remain a mystery to formal reason. Nonformal reason, in contrast, holds out the possibility of offering good, rational counsel for moral problems because properly designed process is central to all moral problem solving, not least when it affects human flourishing, as these vignettes convey.

5.3   Moral Effectiveness: Moving beyond Principles

How can bioethics be effective in handling moral problems? Reports of effective problem-solving from clinical ethicists do not involve the application of moral principles or moral theory.[7] Early in his bioethics career Art Caplan, for example, had two acknowledged successes. One occurred while he was teaching an elective course called "Ethics Rounds" with a psychiatrist and internist in a large urban medical center. The students and instructors would visit patients selected by the students, interview them, and then discuss the moral issues the students and instructors discerned. Once a 90-year-old woman who had no relatives was picked by the students because they were worried about what would happen to her after she was discharged. Caplan dutifully brushed up on his moral philosophy: "I came to the class fully prepared to discourse on theories of distributive justice at a moment's notice, since I rather naively believed the students might benefit greatly from a disquisition on Mill, Rawls, and Nozick in trying to figure out what to do with the old woman" (Caplan 1983: 311). But he quickly got sidetracked:

> As soon as the medical instructors and students had gathered together, we hurriedly set off to find and interview the old woman. We all burst into her room just as she was in the process of defecating. To my surprise, no one was deterred by her behavior, and both the psychiatrist and the internist proceeded immediately

to interview the woman about her life plans, goals, and personal aspirations. I remained uncharacteristically silent during this exchange, and it was only when the class had returned to the confines of the psychiatry lounge to discuss the case that I proffered the opinion that it might have been better to wait until the woman had finished her excretory functions before interviewing her. (311–312)

His reaction "was greeted with some consternation by both students and the other teachers," but they conceded that the privacy of patients is important and should not be overridden by their busy schedules. There was likely more than this at issue (see below), but the value of "applied" bioethics, as Caplan had practiced it, was established: "my esteem among the members of the course was assured for the duration of that particular clinical rotation" (312).

Caplan topped that success when he tackled the thorny problem of rationing scarce medical resources. Hot summer weather made it difficult for persons with emphysema and other respiratory ailments to breathe and drove them to the emergency room for oxygen. But there were only two oxygen units, and throughout the day and night at least a dozen people often were waiting to use them. The staff in the emergency room wanted Caplan to help them develop criteria for fairly allocating their scarce resource.

Again Caplan prepared as an applied ethicist would, by consulting the philosophical and theological literatures on justice and microallocation. He found predictable defenses of allocating scarce resources based on criteria of need, merit, and social utility and of deciding randomly by using a lottery. But he was savvy enough to realize that the medical staff could get that far without either him or the materials he read. Then he had an inspiration:

> It occurred to me that it might be possible to solve the allocation problem by ameliorating the source of the scarcity. I asked some of the emergency room staff if Medicaid/Medicare covered the provision of air conditioners in the homes of persons suffering from respiratory ailments. It turned out, much to everyone's surprise, that the machines could be prescribed and the cost reimbursed. (1983: 312)

Caplan's reputation was secured: "I ascended to the status of moral guru in the emergency room, famous as the man who had solved the oxygen machine crunch" (312).

Neither success resulted from applied ethics, as Caplan recognizes: "… these two cases of efficacious moral action were hardly dependent on analytical rigor or theoretical moral sophistication for their genesis. Indeed, in both cases, ethical theory would have been the wrong place to turn for a solution to the issues under consideration" (312). What Caplan means by "ethical theory" must be understood carefully, however. The ethical theory he has in mind is the norm-constituted brand of ethical theory propounded in philosophy and invoked by applied ethics. Our goal is not to reject ethical theorizing or ethical theory but to develop a broader theoretical understanding of ethics that can account for the kinds of success that Caplan had.

Toward that end, the sources of Caplan's successes should be appreciated. Caplan's first success was the product of observation and critical assessment. In a rudimentary sense, when the students and the instructors entered the patient's room, Caplan saw what everybody else saw. In a more sophisticated sense, Caplan saw what nobody else saw, just as a chess grandmaster can look at the position on the board and assess the outcome, and as an experienced firefighter can look at a burning building and assess the danger. What Caplan observed and the other participants did not was the moral salience of the situation, probably because of his training and experience and because, as an outsider, he was not inured to routine encroachments on patients' privacy, particularly when they are egregious. Moreover, Caplan was willing to expose and challenge the deep-seated assumption that the time of busy doctors, residents, interns, and medical students is more important than the dignity of patients. The consternation he caused suggests that behind the issue laid the deeper problem of the self-interest inherent in treating this elderly woman, or any other patient, as simply a source of information and learning rather than a person in her own right. This prospect speaks to the cultural ethos that governs a hospital, or any other institution, and suggests an inadequacy that is even more deeply embedded and needs to be confronted.

Caplan's second success was the result of creative construction and formal and informal reasoning. Rather than accepting the ineluctability of the scarce resource problem and searching for a way of solving it, he entertained the possibility of dissolving the problem by removing its cause. That is, he stepped outside the confines of the problem as posed and replaced it with a wider problem that admitted a radically different solution, an exemplar of intelligent procedure and a creative act characteristic of nonformal reason. And he proposed that the terms of Medicare and Medicaid reimbursement might be applied directly or extended by analogy to cover the cost of air conditioners for these patients. The sources of Caplan's successes—observation, critical assessment, formal and informal reasoning methods, and creative construction—are the resources of the broader conception of nonformal reason that accounts for the rationality of both science and ethics. And in both cases they led to ethically valuable outcomes that would have been entirely missed had his methods been labeled as irrelevant or nonethical.

5.4 Conclusion

The resources of nonformal reason reveal and account for the rationality of the dying children who imperatively are trying to find out what is wrong with them and what will happen to them. Bringing those resources to the vignettes reveals the limitations of formalist rationality and the notion of autonomy in applied bioethics, limitations that prevent it from providing moral help when moral help is most needed. A formalist account of autonomy as self-determination is acontextual because the locus of autonomy does not matter. Whether it is the autonomy of the individual, the autonomy of the family, or the autonomy of the tribe, the function of autonomy as self-determination is to protect the exercise of autonomy from interference. Moreover, a formalist account of autonomy is empty because it prescinds from the values that are necessary for and relevant to a decision: autonomy as self-determination is in the service of the specific values operative in a context. Formalism does not provide resources for critically assessing and implementing existing values or for formulating and adopting new values. Yet those tasks are precisely what makes the protection of autonomy so essential. Nonformal reason provides the capacity and the resources needed for rationally accomplishing those vital tasks.

II Ethics as Design

6 Ethics as Design and Its Two Distinctive Methods

Re-reasoning ethics begins with an enriched conception of rationality that fits our whole intelligence, not just our logic, and develops a correlatively more encompassing conception of ethics that fits our whole lives. Here we complete our account of how this new rationality re-reasons ethics by presenting Caroline Whitbeck's (1996, 2011) explanation of how design in engineering can be brought to practical problems in ethics and by introducing two distinctive, fundamental methods for rationally designing resolutions of practical problems in ethics: constructing a fully engaged moral compromise and creating a liberated, extended, and enriched wide reflective equilibrium.

Design is used in engineering to work through a practical problem in a practical way. Doing that requires framing a problem, identifying possible solutions, and so on, and improving these across trials, all in the manner discussed in section 4.3. Design in engineering is not just doing that, but figuring out how to do that, and within the constraints applying such as for time and other resources. Doing all that for practical problems in ethics is, as Whitbeck puts it, "doing justice to moral problems" (1996: 9).

Two distinctive methods of problem solving help to do justice to moral problems: compromise and wide reflective equilibrium. Avishai Margalit (2010) recognizes that "compromises are vital for social life." Compromise is widely accepted and utilized in bargaining and negotiation about economic and other matters and in politics, which Margalit calls "macromorality." Compromise also is "center stage in micromorality," the morality of persons and of dealings among individuals:

> We very rarely attain what is first on our list of priorities, either as individuals or as collectives. We are forced by circumstances to settle for much less than what we aspire to. We compromise. We should, I believe, be judged by our compromises more than by our ideals and norms. Ideals may tell us something important about what we would like to be. But compromises tell us who we are. (2010: 5)

Compromise is ubiquitous throughout our moral lives, indeed unavoidable for all finite creatures, yet it can have no place in the logic-based, formalist approach to ethics. The method of fully engaged moral compromise acknowledges the unmet demands of the conflicting ethical claims that are compromised. Moral compromise has to be characterized and clarified, and its central role in ethics defended.

The development of the concepts of narrow reflective equilibrium and then wide reflective equilibrium liberates moral philosophy from the debilitating limitations of formal reason's foundationalism. That was an important advance, but more still needs to be done to free reflective equilibrium from formal reason's perdurable insistence on justification and theory construction. Once fully liberated, wide reflective equilibrium has to be extended to encompass diverse forms of moral deliberation and creative moral options, to accommodate a vast expansion of morally relevant considerations, and to be enriched with compromise as a method for reaching equilibria. The result is a more realistic, more powerful, and rational nonformal ethics.

6.1   Whitbeck on Ethics as Design

Whitbeck defines design problems broadly, as "problems of making (or repairing) things and processes to satisfy wants and needs." A chef designs a holiday meal; a boss designs a work schedule; a scientist designs an experiment. In an engineering school, design courses are taught along with theory courses. Engineering theory courses, Whitbeck explains, "teach students to understand theory and how to apply it to solving problems with mathematically exact and usually uniquely correct solutions." Design courses use "synthetic" as opposed to "analytic" reasoning "to construct good responses to practical problems" (Whitbeck 1996: 10).

What aspects of engineering design, and design in general, are relevant to ethics? Whitbeck describes four and illustrates them with the example of two engineering students assigned the task of designing a child seat that can fit on the top of suitcases being wheeled to an airport gate, can be used on a seat in the airplane, and can be stored either under a seat or in an overhead bin.

The *first* feature is that only rarely, if ever, is there a uniquely correct solution or even a predetermined number of correct solutions. In engineering, Whitbeck reports, "if there is one solution to a design problem, there are usually several" (1996: 10). And both engineering and moral problems can be made simpler, even trivial, when the framing or specification of the problem is so tight that the range of acceptable solutions is small. If we add "must be metal and light" to the child seat requirements, we essentially limit the framing solutions to aluminum plate and tubing. If we want to prescribe what bystanders to an accident should do, the moral options are nicely reduced if we specify that bystanders are capable of making a difference and will not be in danger and that there are no other factors. The moral problem then becomes trivial. Trivial problems selectively reinforce the philosophical penchant for portraying ethics in terms of "doing the right thing," but when there are multiple acceptable solutions, there is no single "right thing to do." Along with providing medical aid, a bystander might, for example, clean up to prevent a fire, warn approaching cars to avoid collisions, phone for an ambulance, and the like, and each might represent a crucial, but competing, constructive intervention.

The *second* feature is that even if there is not a uniquely correct solution, there might be clearly wrong solutions, and among the solutions that are not clearly wrong, some will be better than others. That recognition is a corrective to the glib adoption of the dichotomy that either there is an objectively correct solution or all solutions are arbitrary and subjective, a common presumption promoted by the logical-deduction-from-a priori-principles approach to ethics.

The *third* feature is that when there are two (or more) acceptable solutions, they likely have different advantages and disadvantages and so there is no one demonstrably better (or best) solution. But the differing advantages and disadvantages of multiple solutions might differently fit the different situations of different persons. (That one solution is best for one person while another solution is best for a different person is contextual relativity, not relativism, because the fit of solution to person can itself be critically appraised. The opposition between either acontextual/universal or arbitrary is another false dichotomy encouraged by the logical-deduction-from-a priori-principles approach to ethics.)

The *fourth* feature of design in both engineering and ethics is that solutions must: (a) achieve the desired performance or goal; (b) conform to prescribed specifications or criteria; (c) be reasonably secure against accidents or other untoward events that could produce bad consequences; and (d) be compatible with existing background constraints. These aspects of design are ignored when moral decision making is taken to be the application of a moral theory or principle.

What lessons for design in ethics follow from this analogy with design in engineering? The *first* and most crucial *lesson* is the need to appreciate the rampant ambiguities and uncertainties that are inherent in framing a problem. The engineering students designing a child seat, for example, did not know what features potential users might want or what solutions might already exist, nor did they know how compatible the various requirements for the child seat might be, for instance, whether such a device could accommodate a full range of sizes of preschool children yet be light and easily removable. Problems have to be defined in a manner that preserves as much flexibility as possible. And because often it is not possible to wait for the resolution of ambiguity and uncertainty, fallible but rational judgment is essential.

A *second lesson* from engineering design is that an initial characterization of a problem is only a first step, beyond which lies the devising of possible solutions. As Whitbeck points out:

> [Identifying solutions] is one of many features that distinguish moral problems from formal "decision problems." Decision problems or problems in decision analysis include specifying the alternatives among which one is to decide; that is, a fully defined decision problem is a type of multiple-choice problem. This need to develop possible solutions implies ... that statements of moral problems as open-ended situations do more justice to those problems than do statements of them as multiple-choices. (1996: 13)

Moreover, solutions for the formal decision problems of applied ethics are often binary, for example, dispose of a regulated toxic substance by pouring it down the drain or do not pour it down the drain. But this ignores what the sequelae of these stark options might be. With open-ended, that is, deep problems, framing the problem and identifying potential solutions are interactive, complementary tasks. Whitbeck appreciates that, in order for solutions to be proposed, problems have to be formulated and that "even open-ended statements are only outlines of ethical problems" (1996: 13). Framing real moral problems requires investigating the features and details of a problem and the various contexts within which the problem arises. Only then can promising solutions emerge and be modified or altered as dimensions of the problem shift. With respect to a hypothetical scenario in which her supervisor directs her to dispose of a regulated toxic substance by pouring it down the drain, Whitbeck explains that if that were a real problem, "there would be a particular person who would be my supervisor whose character I might learn more about" as well as "an actual organization (a company, a university) with particular policies that I could investigate" (13). Not to mention investigating other ways of disposing of toxic substances, or of redesigning the experiment so as to avoid their use, and so on.

Whitbeck praises the practice of "brainstorming" because it enables engineers to learn how to develop responses to problems, which is essential to creative construction. But brainstorming, she cautions, "requires an uncritical atmosphere in which people can present half-baked ideas that may later be refined or combined" (Whitbeck 1996: 13). Whitbeck knows, however, that the attempts of moral philosophy students "to probe the complexity of the case are cut off" routinely and peremptorily and that brainstorming is strongly disapproved: "Articulation of any half-baked ideas is discouraged in the many ethics classes where adversarial debate is the primary method used. Although an adversarial format may provide some useful pre-law training, it does not help develop the ability to think constructively about resolving moral problems" (13–14).

Philosophy students are expected to choose between the prescribed options (typically, utilitarianism and Kantianism), to apply the moral theory they deem to be correct, and to defend their position zealously and uncompromisingly. That approach might indeed teach the skills of debating or advocacy, two enterprises where the goal is to win, but it does not teach problem-solving skills, in morality or anywhere else.

Whitbeck provides an example of what she regards as a "rather heroic capacity to brainstorm in the face of criticism" displayed by Amy (Whitbeck 1996: 14), a student who participated in Carol Gilligan's research on moral development (1982). Amy responded to Lawrence Kohlberg's classic Heinz dilemma of whether Heinz should steal a drug that could save the life of his wife but that he cannot afford. Amy, entirely reasonably, recognizes that Heinz's wife should not die if her life can be saved and that Heinz should not steal the drug, and she realizes that if Heinz does steal the drug, he might go to jail and then would be unable to support his wife as she becomes sicker. So, quite reasonably, she explores options such as trying to negotiate with the druggist, borrowing money, and taking out a loan. Amy's creativity is suppressed in a theory-dominated philosophy classroom, and Amy's kinds of options are not available on the multiple-choice questionnaires used in psychology research to determine at which of Kohlberg's Piagetian-inspired six hierarchical stages of moral reasoning a subject is.[1]

In short, with open or deep problems, as Whitbeck explains, framing a problem and devising solutions for a problem are interactive, reciprocal endeavors, not separable, independent activities. Each can inform and direct the other to enhance the understanding of a problem and how best to respond to it. And each constrains the other: conceptions of the problem limit the range of acceptable solutions, and the range of potential solutions limits how the problem can be conceived. The two complementary undertakings have to be integrated in a creative, flexible process of problem formulation and solution. While all of Whitbeck's lessons apply to all design processes, the design literature beyond engineering almost universally highlights the creative exploration of possibilities, both in framing a problem and in identifying potential solutions as the key feature of problem-solving processes.[2]

The *third* lesson is the need to act under time pressure. When time is short, it is important to pursue several possible solutions simultaneously so that alternatives exist if one turns out to be impossible or impracticable. Amy's rejection of the dichotomous choice that is imposed—steal or don't steal—and her ensuing brainstorming produce several options for Heinz to pursue, with various probabilities of success and various timescales for exploring each.

The *fourth* lesson is that problems are dynamic. The nature of a problem can change as circumstances and contexts change, as constraints are investigated and modified or eliminated or new constraints are discovered, as elements of the problem are changed or removed or additional features are introduced, and as progress is made by addressing some aspects of the problem. And such changes can also change the conceptions of what a good solution might look like and what methods might be appropriate to uncovering solution(s). All these elements can coevolve as potential whole or partial solutions are explored. This dynamic fluidity is, in fact, the general nature of design process. Moral decision making should not be portrayed as a matter of solving static, completely but narrowly defined moral problems by choosing one of a limited number of static, completely but narrowly defined options. A moral problem, as Whitbeck trenchantly puts it, is not a "math problem with human beings" (Whitbeck 1996: 15).

Michael van Manen and Nicole Kain (2017) provide a gripping example of how ethics as design can work. It concerns the creation of a policy that would permit the donation of organs for transplantation, particularly hearts, from infants who die in a neonatal intensive care unit (NICU). As an illustration of the staff's commitment to improving care, van Manen and Kain describe a previous redesigning of the NICU. Its open-ward structure was altered to create single-family rooms that would foster family-centered care and provide comfort for both the infants and their parents and other family members. When their families were not in the NICU, though, the babies were often isolated in the single-family rooms. So the NICU was further redesigned—the walls between adjacent rooms were made collapsible so that staff members could interact with infants when their parents were absent. Both material redesigns of the NICU help to engage and assuage the infants and their families. Similarly, designing and implementing a policy that would allow organs from infants who die in a NICU to be donated for transplantation would prolong the lives of other infants who are in a NICU. Van Manen and Kain appreciate that a policy has to "reflect thoughtful intent, careful deliberation" (2017: 20). For that outcome to be attained, the process of developing the policy has to be rigorous and transparent, and that means the process has to be broadly participatory and iterative. The creative design of their process engaged the three activities of "*exploring, ideating*, and *testing* carried out in iterative fashion" (23; emphasis in original). The goals of exploring are to "expand upon a specific problem and existent solutions," to "find patterns and exceptions," and to "identify limitations and boundaries" (23–24). The goals of ideating are to "redefine problems, creating space for novel standpoints, alongside practices such as sketching, drafting, and modeling imagined possibilities for design." Testing is experimenting by addressing "instrumental, technical, aesthetic, and ethical aspects of design in the form of prototyping and implementation." Testing requires that the process be robustly iterative: "design inherently necessitates re-design because a solution for a given problem may actually change the nature of a design problem itself" (van Manen and Kain 2017: 24).

The work of developing the policy was done by an advisory group and the research staff, comprised of van Manen and Kain—a neonatologist with experience in qualitative research and a research assistant with experience in nursing and public health. The research staff supported the advisory group, which was comprised of 18 members who were "expert" in the care of NICU patient families and had diverse backgrounds, experiences, and views about donation after circulatory death (DCD). The advisory group adopted a focus group method because the "potential 'novelty' that emerges from focus group discussions is congruent with the creative crux of design" (van Manen and Kain 2017: 24). The advisory group held nine audio-recorded meetings that examined issues about "the processes surrounding DCD: eligibility criteria for DCD, consent process for DCD, pre-mortem interventions for the DCD donor, palliative care for the DCD donor-family, declaring death of the DCD donor, and post-mortem care for the DCD donor-family" (25). The content from the meetings was turned into four Survey-Briefings that were presented to "neonatologists, nurse practitioners, bedside and specialized nurses, neonatal subspecialty trainees, respiratory therapists, dieticians, pharmacists, and social workers" and circulated to parents who had had a child in the NICU. Subsequently a DCD policy was drafted and then "submitted to hospital administration, legal representatives, and other external experts and end-users" for their assessments. (29). Their reactions are likely to determine whether a policy is implemented. Regardless of the outcome, the depth and the breadth, the content and the design of this policy-making endeavor render it a model of nonformal reason. (See also the NICE procedure, discussed in chapter 9.)

Rather than being depicted as conflicts, moral problems are problems that are multiply constrained. Comprehensive critical assessment is needed to determine which constraints might be satisfied simultaneously and to what extent each might be satisfied, and which constraints are impossible or unlikely to be satisfied. Creative construction, operating in tandem with comprehensive critical assessment, will search for innovative ways of satisfying various combinations of the constraints to varying degrees. This characterization already suggests a potentially rich space of alternatives for searching. And even when attractive solutions might exist, there could be ways of improving them. What is essential for solving problems in ethics, whether individual or social, is, as in engineering, designing a creative problem-solving process that is open, sensitive to process and timing, flexible, and responsive. Nonformal reason provides the resources that can rationalize processes for investigating, discussing, and resolving moral problems in this way. Whitbeck's characterization of moral problem solving as akin to engineering design provides the right way to conceptualize moral problems and to understand how nonformal reason provides the resources to deal with them. And whereas engineering problems are replete with potential and actual compromises, and practical ethics problems likewise, so that it is an essential part of rational method for problem solving, compromise has no place in formal moral philosophy.

6.2 The Deduction-from-Principles Account of Compromise: A Critique

There are two logically respectable ways of resolving moral conflicts. One is to revise the applicable obligations that are inconsistent to make them consistent and then deduce what ought to be done from the resulting set of consistent obligations. The other is to identify a consistent set of obligations that is deemed to be more compelling in the particular circumstances and then deduce what ought to be done from the preferred set of more weighty obligations. These two strategies, described respectively as specification and balancing, are adopted in the current practice of moral philosophy for resolving conflicts among principles. Insofar as the result of either method can be called a compromise (a matter discussed later), it is a conflict-removing compromise. Conflicts must be expunged because they are logically intolerable. We argue that this orthodox philosophical approach to moral conflicts, and whatever conception of compromise it might yield, are not an accurate representation of either the nature of, or the resolution of, most ethical dilemmas, nor do they support our moral conceptions of a morally mature person.

Specification[3] renders applicable principles consistent by modifying the formulations of the principles to reduce their ambit or scope (in logical terms, the extensions of the principles). The statement of one or more of the conflicting principles is amended to restrict its applicability so that the principles no longer conflict—for instance, in the way the conflict between relieving suffering and preserving life in Ms. B's case (see section 5.2.1) was removed by restricting the latter to nonterminal conditions. Because the ensuing restricted obligations that apply now apply fully, there is no conflict, and what ought to be done is clear. But how is this conflict-removing restriction to be understood and justified? That depends on the underlying epistemology. With foundationalism, if the restrictions imposed can be justified by deducing them from more basic ethical principles, conflicts can be resolved without the need for any extralogical reasoning. If not, the only resort is to an a posteriori discovery of an a priori truth—to claim that the newly restricted statement of the principle simply corrects a hitherto mistaken formulation of the principle and reveals what the true content of the principle always has been. With coherentism, the only option is to invoke a notion of coherence or fit, with reflective equilibrium being the preferred philosophical candidate. We consider these options below, but first turn to the other familiar strategy, balancing or weighing.

Balancing or weighing is the approach of moral intuitionism. Applicable principles that conflict vie to be the sole principle, or priority principle, to be applied. Principles are regarded as being prima facie reasons for a decision, and intuition is used to "balance" or "weigh" inconsistent prima facie principles to determine which should prevail in the circumstances. Balancing or weighing eliminates logical inconsistency by rejecting the application of the prima facie principles deemed weaker in their particular contests with other morally relevant principles.

Both the strategy of specification and the strategy of balancing or weighing encounter three problems. The *first* problem is how to provide a clear sense of rational justification for the favored resolutions of conflicts. Consider first a foundationalist approach. That there is a problem is obvious for balancing and weighing because those techniques appeal solely to intuition, but intuitions, however they are understood, are notoriously variable across agents and modifiable across experiences. Given that diversity, how are conflicts between inconsistent intuitions to be rationally resolved? The intuitionist has no response. The specificationist who appeals to a posteriori discovery of a priori truth is on no better ground: the mere insistence that the newly restricted formulation of a principle simply represents the principle as it always has been is no more rationally compelling than an intuition. The claim that restrictions can be deduced from more basic principles at least offers the possibility of a rational justification. Its problem, however, is that providing those deductions has proven to be very difficult, if not impossible. Deductive rationality is stuck with consistency and inconsistency as its only relations among principles, so on what grounds could one moral principle be shown to give in to, or make concessions to, another moral principle? The only recourse would seem to be an appeal to second-order moral principles that prescribe the applicability of first-order moral principles, but such principles are neither readily forthcoming nor have much moral force. Indeed, any such second-order principles seem post hoc, presupposing and endorsing the hard work of moral consideration rather than enlightening it.

A coherentist approach has the opposite problem. Coherence is readily attainable when it means no more than logical consistency, but logical consistency is too weak to provide any meaningful justification of restricted principles—any two completely unrelated propositions are logically consistent. So consistency could be attained simply by adjoining any consistent restrictions to the conflicting principles. Recognizing that, Henry Richardson (1990) proposes that it is possible to justify restricted principles by incorporating them into a coherent moral theory so that they fit together with other such principles. But again, the notions of "coherence" and "fit" remain refractorily vague and nebulous. Intelligible coherentist consistency is easily attainable, but its weakness provides no meaningful rational justification.[4] In sum, within traditional moral philosophy there are no readily evident ways out of this first problem.

The *second*, and related, problem is that without a rational justification for the restriction or intuitive dominance of a principle, the parties to a compromise are left with their moral integrity violated.

The *third* problem is the most revealing: both specification and intuitive dominance fail to capture the ongoing claims of unsatisfied moral obligations, which both characterize our moral condition and drive our moral development. For instance, they discount the struggle of the single parent, where both obligations (to a child and to work) continue to bite and foster regret,[5] and similarly fail to do justice to the many other intractable struggles at the core of our moral lives.[6]

This third problem also exposes the fundamental conundrum for the notion of conflict-removing compromise: How can a conflict-removing compromise be a compromise?[7] If there is no conflict, there is nothing to compromise. So where might any putative compromise be located? Specification produces a consistent set of restricted principles as outcome, so no compromise can lie there. Perhaps the compromise lies in the restrictions that curtail the obligations. But it cannot be there either if the restrictions are to be justified as really always having been there, and especially if they are justified by deduction from more basic ethical principles, since the restrictions then would be a fundamental part of the principle and not a compromise of it. Restrictions justified in this way would render compromise no more than an initial illusion that the principles were unrestricted. The same holds for intuitive balancing and weighing. Compromise cannot lie in the outcome because that is a single dominant principle, and it cannot lie in the restrictions if these are to have moral validity, since they have to be eternal moral truths discovered by intuition. Compromise again becomes no more than an initial illusion that the principles were unrestricted.

In sum, while conflict-removing consistency provides a basis for moral action, vindicating the validity or rationality of the outcomes of specification or intuitionism eludes standard moral philosophy. Establishing the rationality of compromise requires a richer understanding than this of what compromise is and how compromises are formed and where they lead. It requires the realization that recognizing, and a fortiori creating, an intelligent genuine compromise is a matter of skilled judgment formed and guided by nonformal reason. This realization—and radical departure from orthodox moral philosophy—sets the stage for reconceiving the nature of compromise and its role in moral life, along with reconceiving the conception of rationality that underlies compromise.

6.3 Moral Compromise

Compromise is a pervasive fact of life.[8] It is occasioned when the several obligations we acknowledge in a situation cannot all be fully met and we cannot avoid the conflict because inaction will generate a worse outcome. Compromise occurs, for example, when a single parent cannot both satisfactorily parent a young child and earn the income that living together demands; when a doctor cannot both preserve a patient's life and control pain or respect a wish to die. War zones are permanent sites of wrenching moral compromises, but they differ from political or commercial arenas only in degree, not in kind. For people striving to lead moral lives, compromise is always unsettling, and it can be devastating. Yet compromise can also be uplifting. Think of the single mother who determinedly and skillfully picks her way between caring and working, committed to both and faithfully bearing the extra effort of feeling and thoughtfulness. She is an exemplar of moral and emotional maturity, someone who has grown, not shrunk, through the experience of compromising (without in any way reducing her preference to eliminate the compromise).

Would-be conflicts that are removable do not involve compromises. Suppose a doctor, faced with an aged patient in extreme pain, concludes that the patient can be made sufficiently comfortable to satisfy the obligation to control pain without the indicated high doses of a drug that would accelerate the demise of the patient. Then there is no need to compromise; each obligation can be satisfied. Or suppose the doctor concludes that the obligation to preserve life does not apply to terminally ill patients for whom death is imminent but only to patients who have the prospect of continued life. Here, too, there is no need to compromise; one obligation does not apply, and the other is fully met. But this is not the situation of the single parent. The obligation to bond with and nurture the infant is ongoing and pressing but is practically defeated if there is no bed to sleep in, no food on the table, no time to play. Likewise, the obligation to earn a living is ongoing and pressing but is defeated if she is never with her child. Neither can be set aside, and neither can be fully met. Moreover, when each is met to the extent it can be, the demands of the fully applicable but partially unmet obligations continue and will be keenly felt. The parent remains acutely aware of the time not spent with the infant and the needs and advantages of life not provided, as well as of the career opportunities foregone. This is the stuff of real, conflict-preserving compromise because the forces of the competing obligations remain undiminished or unrestricted throughout, yet each is partially defeated in the outcome, to a degree deemed proportionate to its proper role in the current circumstances.

It might be tempting to argue that when individuals inherently cannot satisfy all the obligations they recognize, then, since "cannot" implies "not ought" (since "ought" implies "can"), they are not under an obligation to do what they inherently cannot do. That is so, but it does not release them from compromise. Although they are not obligated to *act* so as to meet all their acknowledged obligations, it is quite consistent with that constraint that they *are* obligated to continue to acknowledge (to "feel") the press of all the obligations on them. That is in fact what happens, and it is crucial to our being a moral person that it should do so, for it keeps us engaged with the entire moral order. And that engagement in turn provides the widest basis for appreciating new options for improving the situation as they arise. (See also section 6.3.2 below.) If we are not to abandon these unmet obligations to some limbo of unacknowledged moral experience, it is important to acknowledge a legitimate place for compromise in our lives.

In doing so we should not confuse two distinct kinds of "compromise." There are compromises of principle that are immoral, for instance, accepting a bribe to allow an illegal act. If a proffered resolution of a conflict among obligations were to include such a compromise, moral integrity would be violated. Those are corrupting compromises. But that problem is quite distinct from the issue of whether conflict-preserving compromises per se are immoral or jeopardize moral integrity. We must hope not, since they stand at the core of our moral experience and development and are ubiquitous, found not only in every private life but also in every public policy, every design process, and so on. The goal, then, is to find a proper role for noncorrupting compromise in the moral life and a proper rationale for that role.

6.3.1   The Nature of Compromise

Martin Benjamin provides an instructive example of compromise at work in a moral conflict in an intensive care unit (ICU) (Benjamin 1990, chapter 2). A young, single woman who has suffered severe brain damage from viral encephalitis has been admitted to the ICU, and her overwhelmed parents have ceded decision making to the medical staff. An experienced critical care nurse and the attending physician in the unit disagree about whether aggressive treatment should be continued. Other members of the staff also disagree about how the young woman should be treated. The head nurse argues that it is highly unlikely that she would want to be kept alive in these circumstances, that even if with treatment she would live, she would be likely to have multiple, severe, debilitating deficits, and that continuing aggressive treatment is very expensive and not worth the cost given the exceedingly low probability of obtaining significant benefits when other patients could benefit significantly from those resources. The physician believes that the young age of the woman, the sudden onset of the disease, and the patient's previously excellent health suggest that if anyone could recover, she could, and that recovery is of most benefit to the young. The physician also appeals to the inherent value of human life and the importance of remaining steadfastly dedicated to preserving and prolonging human life.

Because the nurse and the attending physician are at an impasse and the staff is divided, the status quo—aggressive care—continues. For many reasons, including putting an end to a distressing, demoralizing stalemate, a resolution is needed, perhaps a compromise. Benjamin proposes an account of compromise that has three components: an explanation of the circumstances in which searching for a compromise is viable; a description of the process of compromise; and an analysis of what is required for the outcome of a process of compromise to be a genuine, or strict, compromise. *The circumstances of compromise* consist of factual uncertainty; moral complexity; a continuing, cooperative relationship between the parties; an impending, nondeferrable decision affecting both parties; and scarcity of resources. The more of those features that are present, the stronger the case for a compromise. Whether to pursue a compromise is a matter for nonformal judgment. *A process of compromise* includes an exploratory discussion among the differing parties, marked by mutual respect, a fair hearing of argument, and a willingness on both sides to respond to "new information, insights, arguments, or understanding" (Benjamin 1990: 32–33). *A genuine compromise* results when substantive differences between the parties continue, so that for there to be a resolution the parties have to agree to split their differences, each side giving up something. For instance, the parties could agree to continue aggressive treatment but only for a specified time, after which they would assess the patient's condition using criteria they have mutually accepted. If by that tine the patient's condition was improved, aggressive treatment would continue. If it remained the same or deteriorated, less aggressive treatment would be instituted and more importance would be given to the equitable and efficient use of the resources for treating other patients (34–35). This is a genuine compromise: although agreeing to the outcome, does the nurse still regret the other patients' losses of support, and does the doctor lament the violation of the obligation to vigorously pursue life? Surely yes and yes. Compromise is the reluctant bending to the strictures of finitude in the face of the continuing matrix of acknowledged obligations. There is no action that can be taken by finite agents in this world, individually or in a group, that can satisfy all our values, and so each act is in principle a conflict-preserving compromise, even if a minor one in the wider compass of life.

6.3.2 The Moral Life of Finite, Developing Agents

Finitude means that we begin in ignorance—ignorance not only of the substance of a good life, but of the terms in which to conceptualize it and the methods to use to validly acquire it—and that we possess limited resources, including reasoning capacity, with which to pursue relief from these limits. For finite agents, life is full of conflicts among valuables, whether those valuables are goods all of which cannot be obtained or obligations all of which cannot be fully satisfied. Compromise derives from our finitude in the face of the complex, multifaceted character of valuing. To return to our earlier example, single parents properly value both caring for their children and working, but their finitude prevents them from fully satisfying the obligations of both. This conflict is contingent insofar as a parent's need to work could be removed by a benefactor (governmental or personal) or a lucky accident (winning a lottery), which makes clear how finite monetary resources underwrite it. But this feature alone misses the intrinsic value to a parent-as-social-participant of working, developing work-related skills, taking pride in occupational accomplishments and so on, and the equivalent social dimensions to parenting (developing play-group skills, and so on). In this regard the finitude constraint is more fundamental: finite creatures must learn by social interaction at a specific time and place, but caring and working, then, are at least attention-competing, and typically spatiotemporally mutually exclusive, activities. Similarly, a nurse's capacity to care remains finitely distributed among patients even with expanded resources, and to propose increased staffing as a resolution only promotes conflicts about the allocation of finite resources to a broader social context.[9]

Coping with moral ignorance and conflict is the essence of the moral life of finite agents, and Betty Winslow and Gerald Winslow astutely identify the implications and the dangers of this plight:

> To the extent that two or more of our own firmly held moral values are in conflict, we know for certain that any decision we make will represent a loss of something important to us. We also know that we could be wrong. The complexities and the uncertainties of the matter render us less than completely confident about our choices. This acknowledgment is nothing more (or less) than the appropriate level of humility, given human limitations. Such humility helps us to remember the value that was lost when its competitor was given priority. It is often tempting, of course, to make the moral life simpler than it really is by pretending that one of the values in conflict is not so valuable after all. But this approach to eliminating moral perplexity is neither honest nor prudent. (Winslow and Winslow 1991: 318)

Here again the abstract simplifications of moral philosophy are not faithful to our moral lives. Rather than representing a minor and unfortunate aspect of life to be eradicated, as the logical consistency approach would have it, conflicts among values stand at the core of our moral lives, since there is nothing of moral substance a finite agent can substantially improve without resorting to compromise.[10]

At least as importantly, living with conflict-preserving moral compromise is what drives our moral development. As we noted earlier, the single mother who effortfully navigates between caring and working, and willingly bears that effort, exemplifies moral and emotional maturity, having grown in the process of compromising. Moreover, the continued conflict of values means that she remains alert for opportunities to reduce the remaining conflict, for example, by acquiring a job with on-site child care facilities or child-friendly hours, and/or co-opting a family member to assist in caring, and/or campaigning for such employment and other improved sole-parent support. Both as individuals and as a species we have to work our way laboriously toward improved moral knowledge and practices; that is one of our greatest developmental tasks.[11] Moral development proceeds substantially through the experience of conflict resolution—it is our key gateway to moral improvement.[12]

We have already noted that the ongoing matrix of acknowledged obligations leaves us permanently open to revisiting and improving our past decisions in the light of our accumulating experience with relevant succeeding decisions. But there is another aspect also worth noting: the construction of a richer coherence. The continuing pressure of the full range of acknowledged obligations ensures that creative construction will search for the richest ways possible to do justice to all obligations, rather than just those obligations met by the momentary compromises we are forced to make. This role of creative construction is crucial in overcoming the fragmented normative myopia that might otherwise ensue from the trail of constrained practical actions we must take. In chapter 9 we provide an extended case study of a prominent health care policy issue, the allocation of scarce kidneys from deceased donors for transplantation, where just this process plays out in a surprising way that can only be understood as a compromise. It is the history of improved resolutions to problems of compromise, typically through ways that lessen their constraints, that gives force to our guiding idea of ethics as design for flourishing.

Finally, the continuing pressure of the full range of acknowledged obligations provides the proper basis for extending and enriching the compromise policies we have. The full panoply of norms in force orients us normatively to new situations; we enter a situation alert for how our compromised but still present values might play out in the new circumstances. It is the continued appreciation of the undiminished obligations that guides the adaptation of solutions to the particularities of other contexts. For example, because of the exploration that the women who received genetic counseling did with respect to their families' core values, adaptive capacities, and, yes, limitations on these as well, they were equipped to respond to a wide range of challenges to the futures of their families, from caring for variously handicapped offspring to moving to a new country. As situations change, the claims of some values will be more easily or comprehensively obtainable, while the claims of others will recede—for example, because of higher associated costs—so compromises among them will be re-struck. Doing that renders ethical decisions context dependent. This conception of a normatively oriented agent, forced to continually compromise while accumulatively learning through that ongoing process what is possible and rationally preferable across the terrain of practical situations, provides exactly what is needed to understand the developmental process of becoming both a mature person and a wise actor.[13]

From this developmental perspective conflict-preserving compromise is in itself no assault on integrity, as Benjamin feared it would be. To require that conflicts be removed is to set the bar for moral performance above what is in principle achievable for finite agents, which is incoherent. And, we have insisted, it would be to miss the real moral quality to be had through struggling with compromises. Indeed, a person would be considered to lack integrity if, striving to reduce the stress of unsatisfied acknowledged obligations, they erected the fiction that for each conflicting pair of obligations one did not apply. Moral integrity is not the mere preservation of consistency. Moral integrity is an ideal to be pursued, the virtue that, in their discussion of compromise and integrity in nursing ethics, Winslow and Winslow describe as "constant sincerity of moral purpose" (1991: 309).

As finite agents we are perpetually caught in moral conflicts, and the best we can ever do is satisfy obligations within our available resources and to the degree they are important in the specific situations in which they arise. For a finite agent, doing that precisely *is* having integrity. Virtue *is* learning to judge wisely which obligations are relevant, what their comparative importance is in a specific context, and which kinds of compromise are best overall. We do not even see any a priori case for inherent patterns of priority among conflicting general obligations, as the deduction-from-principles approach might like. Life may just present us with a set of situations wherein every combination of priorities appears somewhere in best solutions—and yet we can have moral integrity. Thus we conclude that conflict, and how it is handled, especially in compromise, should be embraced as being central to identity and moral integrity.

In this regard it is instructive to review Benjamin's struggles over moral integrity. Benjamin holds that a conflict-preserving compromise threatens moral integrity because the parties to the compromise agree to act in a way that does not fully uphold their obligations. If both the nurse and the doctor regard the outcome of the ICU process as superior to their initial positions and adopt it for that reason, there is no compromise and no worry about their integrity having been violated. But if they continue to differ, for example, about the relative weights to be attached to saving life at all cost, the probability of success, and the fair and effective allocation of scarce resources, and remain convinced of the correctness of their initial positions and agree to the outcome only as a practical, expedient way of ending a disagreement that has to be ended, then there is a worry about compromising their integrity, not just their particular moral principles. As Benjamin puts it, "moral wholeness or integrity, it would seem, is incompatible with their agreeing to ... a compromise outcome" (1990: 36).

His response to this worry is, in the ICU case, to broaden the terms in which the problem is framed to include the participants' professional obligations. The enveloping issue then is how the parties to the disagreement can manage their disagreement in a mutually respectful manner that will allow them to maintain amicable relationships and continue to work together productively in the ICU and, at the same time, emerge with their personal integrity intact. Benjamin quotes a passage from Arthur Kuflik to make this point:

> [T]here is more to be considered than the issue itself—for example, the importance of peace, the presumption against settling matters by force, the intrinsic good of participating in a process in which each side must hear the other side out and try to see matters from the other's point of view, the extent to which the matter does admit reasonable differences of opinion, the significance of a settlement in which each party feels assured of the other's respect for its own seriousness and sincerity in the matter. (Kuflik 1979: 51, quoted in Benjamin 1990: 36)

In this regard the problem for the nurse and the doctor in the ICU is like the problem for K'aila's parents (section 5.2.3): they are deciding not just for someone who cannot decide for himself or herself but for themselves as well. The problem must be framed in a way that includes that crucial dimension.

Benjamin believes that framing the problem more broadly defuses the worry about integrity. The kinds of considerations that Kuflik identifies, he points out,

> reflect values and principles that many of us hold dear and that partially determine who we are and what we stand for. If, therefore, we suppose that ... [the nurse and the doctor] also place a high value on tolerance and mutual respect, it is not so clear that agreeing to the proposed compromise constitutes a threat to their integrity. On the contrary, taking into consideration all of their values and principles plus the fact that they disagree and that they are in circumstances of compromise, the compromise solution may be for them more integrity preserving than any available alternative. (Benjamin 1990: 36–37)

If the nurse or doctor is troubled about compromising her principles in deciding what should be done for the patient, she can assuage that concern by realizing that she is not compromising her principles about what kind of person she should be and how she should handle disagreements with others. She remains open-minded, fair, respectful, and tolerant.

This reformulation of the problem does not, however, remove the threat to integrity that compromise poses. Benjamin is right to introduce considerations about what kind of person the nurse and doctor ought to be because doing so commendably recognizes that moral problems are people's problems, and he is right that this compromise does not violate the integrity of the nurse or doctor, but he is wrong about why the compromise does not violate their integrity. As morally relevant as the new considerations are, they do not, as Benjamin supposes, eliminate the threat to integrity. Rather, they shift the locus of the compromise. The two parties now have to make a comparative judgment about how important their obligations to the patient are and how important their obligations to themselves are. If the initial conflict-preserving compromise about the good of the patient jeopardized their moral integrity, so does an expanded conflict-preserving compromise about the good of the patient and their own good.

Benjamin's myopia likely has the same source as his fear about preserving integrity, namely, his tacit assumption of the deduction-from-principles account of morality, where the logical inconsistency of principles brings moral incoherence. Recognizing that morality is much more expansive and much more complex than reasoning from principles is what removes the threat to integrity. Indeed, one might worry more about the integrity of those who too easily claim to reach consistent positions for too easily avoiding conflicts and the labor of conducting themselves with unceasing fidelity to all the dimensions of their acknowledged obligations.

The compromise between the nurse and the doctor encompasses the full panoply of relevant moral considerations and the continuing salience of the unfulfilled obligations. The two parties will regret what they ought to do but cannot do for this patient and their other patients, and they will regret the respects in which they ought to be but cannot be better persons. At the same time, their regret will give them the awareness and the motivation needed to search for ways to avoid or mitigate similar conflicts in the future—to improve what they do and how they do it for the benefit of the patients and the staff in the intensive care unit.[14] By continuing to acknowledge the entire matrix of obligations, despite having to compromise in acting, persons retain their full normative orientation to the world and their full awareness of the compromise made, and so they develop

morally and are continually primed to do better in the future as circumstances, including self-made circumstances, permit.

6.3.3   The Rationality of Compromise

Resolutions of moral problems will not preserve moral integrity if they
are irrational. But how can it ever be rational to compromise basic values?
When there is no morally preferable and practicable alternative, as in the
ICU case and as happens often in our lives, how then do we go about rationally constructing and accepting a compromise? The model and practice of
nonformal reason presented in chapters 4 and 5 provide the resources, and
the ICU case is an instructive example.

Utilizing nonformal reason in a process of compromise involves the joint
employment of its four resources—observation, formal and informal reasoning methods, creative construction, and comprehensive critical assessment. Benjamin's account of the ICU process, though brief, clearly implies
the use of these tools, from the clinical observation of patients and the
pragmatic observation of resource scarcity, through the creative construction of anticipative scenarios for the competing uses of the scarce resources,
the reasoning about the likely satisfaction of values in those scenarios, and
the critical analyses of the ways the scenarios compete; from which among
possible compromises are leveraged, to the critical, reflective review of the
overall situation to ensure clarity, comprehensiveness, and lack of bias.
How all these resources are brought to bear in a detailed and mutually integrated way cannot be reduced to any simple procedural rules or specified in
advance; their effective and productive use is a matter of the wise design of
a process for the circumstances, and that too requires judgment.[15]

In particular, there are many places in the ICU deliberation process for
pursuing reflective judgment. It can be used to establish the scope and
nature of the problem, for example, whether the disagreement requires a
compromise and, if so, where it lies on a spectrum of relations from con-
flicted to cooperative among the participants and what values are implicated, not only in the focal situation—providing treatment to the ICU
patients—but in the relationships among the participants in the decision-
making, the members of the ICU staff. A mature appreciation of a problem
can emerge only after substantial discussion of all the various aspects of
the situation. In the ICU case that examination includes the clinical prognoses of the ICU patients and the best treatments for them, the available
resources and whether they might be feasibly augmented and, if so, at what
cost, the likely consequences of compromising the indicated treatment for
each patient, and so on. Then questions of which treatment scenarios are

most salient, what values are at stake in each scenario, and how these values might be "traded off" have to be addressed. Each of these matters is subtly complex, and the three are interrelated. The aptness and thoroughness of their consideration contributes centrally to the rationality of the overall process.

A process that is wide enough to encompass the range of relationships involved in practical moral decision making, and which incorporates MEU as simplified sub-cases as appropriate, is wide reflective equilibria (WRE). The notion of narrow reflective equilibrium adheres to logical processes alone to achieve integration, so it needs to be transformed into a reflective process that utilizes the resources of nonformal reason to create rich reflective practices capable of supporting an innovative problem-solving process. This is the subject of the next section.

6.4   Reflective Equilibrium Liberated, Extended, and Enriched

John Rawls's initial formulation of a decision procedure for ethics (1951) and his subsequent conceptions of a narrow reflective equilibrium and a wide reflective equilibrium (1971, 1974) provide an introduction to, and motivation for, liberating reflective equilibrium from not just formal reason's foundationalism but formal reason's demand for justifications of right answers,[16] by extending the width of reflective equilibrium and enriching reflective equilibrium to encompass moral compromise. Extracted from the prevailing philosophical and epistemological background, reflective equilibrium can focus on the learning process rather than the features of its outcome product, in particular it can employ the resources of nonformal reason. The resulting new regulatory conception of reflective equilibrium has the resources to be far more powerful and far more helpful. And it can accommodate, as a formally degenerate version, formal logic-based philosophy's account of reflective equilibrium. In this section we examine wide reflective equilibrium (WRE), now embedded in nonformal reason, as a key tool in rational ethical decision making.

6.4.1   Reflective Equilibrium

At its root equilibrium is a stasis among entities that pull in different directions, whether those entities are literal physical forces or metaphorical social pressures or normative moral obligations, with the additional condition that any movement away from the stasis creates a restoring force that returns the system to it. A simple physical case is that of a marble at the bottom of a round-bottomed bowl: the marble is at rest in the center of the bowl's bottom, and any movement away from there involves climbing up the side of the bowl whereupon gravity returns the marble to the bottom. In this case the marble at equilibrium is motionless and the equilibrium is static, but there can also be dynamic equilibria in processes where a stasis in some features of the process can be restored if disturbed and all the while the process continues, indeed its continuing is typically what provides the restorative action. Standing waves in a river rapids provide a simple physical example. Almost all the physical equilibria of functional interest in living systems are dynamic equilibria that leave the underlying system processes (metabolism, memory updating, and so on) running.

Equilibria in social-pressure metaphors are dynamic. With political equilibria, for example, all the underlying social processes, from bread baking to zoo keeping, are typically presumed to be running while, for instance, opposing parties may reach an equilibrium formulation of an education policy such that all options to depart from that stasis generate net incentives to return to it. Similarly for equilibrium among moral obligations. All the other aspects of personal and social life are presumed to be proceeding in a context while an agent locates an option from which there is no net reason to depart and thus reaches an equilibrium in their commitments. In arriving there the agent has carefully considered each of the countervailing obligatory considerations—that is, the values whose pursuit would, taken separately, move the agent away from that option—with respect to all the relevant consequences and manners of trade-offs among them available in the context. Hence the term *reflective* equilibrium. Reason demands reflective equilibria for all human commitment equilibria, individual or social, since anything less implies a relevant factor that has not been taken into account, and this violates at least the systematic critical assessment requirement of nonformal rationality. Of course, equilibria hold only as long as their inputs hold. Changes to a moral situation—for instance, new information that alters ethically relevant characteristics, or new values that are injected from legislation or from new institutional responsibilities—will reopen the matter, with the present equilibrium often leading to a new reflective equilibrium.

6.4.2 Wide Reflective Equilibrium

The decisions of Ms. F about her death (chapter 2), of the ICU staff about continuing aggressive treatment for a young encephalitis patient (section 6.3.1), of the women at risk of having a child with a genetic condition (chapter 1), and of the Kidney Transplantation Committee about the allocation of kidneys from deceased donors for transplantation (chapter 9) affect different numbers and ranges of people, but that variability is not itself ethically significant and is not what we mean by the width of equilibrium. Rawls's initial method of reflective equilibrium worked back and forth between considered moral judgments and principles to achieve coherence between the two. That narrow reflective equilibrium evolved into a wide reflective equilibrium for justifying a moral theory by reaching coherence among a much wider array of considerations in the same back-and-forth manner between considered commitments. Norman Daniels explains that wide reflective equilibrium is superior to narrow reflective equilibrium because it marshals "the broadest evidence and critical scrutiny," encompassing

> beliefs about particular cases; about rules, principles, and virtues and how to apply or act on them; about the right-making properties of actions, policies, and institutions; about the conflict between consequentialist and deontological views; about partiality and impartiality and the moral point of view; about motivation, moral development, strains of moral commitment, and the limits of ethics; about the nature of persons; about the role or function of ethics in our lives; about the implications of game theory, decision theory, and accounts of rationality for morality; about human psychology, sociology, and political and economic behavior; about the ways we should reply to moral skepticism and moral disagreement; and about moral justification itself. (Daniels 1996: 6)

The core idea of WRE is that the greater the range and diversity of the considerations that underlie a particular reflective equilibrium, and the more those considerations mutually complement and support one another, the more reasonable it is to rely on that reflective equilibrium as a rational basis for action.

The kinds of beliefs Daniels adduces pertain more to moral theory and moral philosophy, however. Developing WRE for a particular moral problem is a somewhat different undertaking. In that regard, return to the ICU staff's disagreement about continuing aggressive treatment for the young woman with encephalitis. Complementing the ICU staff with the staff of an IDU (infectious diseases unit) would be of little ethical significance; it would add numbers but would not expand the relevant medical expertise or ethical experience. We already have seen, though, how diversity between doctor and nurse in this case can significantly test the ethical options, as we have seen the diversity at work in addressing the plights of the children dying from leukemia (section 5.1). The different roles of doctor and nurse provide different insights into patient care and generate different priorities among values, perhaps even some different values. Together the wise judgments proceeding from both roles provide a more nuanced and critically tested exploration of options.

Another relevant diversity worth pursuing would be to search the ICU for occasions when trade-offs have been made between the values of preserving a young life and maximizing quality-adjusted life-years (QALYs) across other patients, to see what considerations were entertained, what arguments were offered, and how and why decisions were made, and whether there were any pertinent specific differences in those situations.[17] Other trade-offs between one and many, young and old, and high risk–high reward and low risk–low reward might also be consulted. And perhaps especially relevant, past compromises reached by the ICU staff (other than the temporarily-support-then-reassess sort) could be sought, and, if found, assessed for their appropriateness and their ethical and medical suitability in the current case. If agreement on a course of action in the ICU emerges from such a network of investigations, the connections and similarities discovered will mitigate the distress of the staff and perhaps strengthen an evolving WRE about a clinical and moral approach to troubling ICU decisions. Alternatively, the outcome might convince the ICU staff that in some circumstances they should not adopt practices from elsewhere. Nevertheless, with a risky or innovative decision, wider coherences could possess increased value because they provide the most similar guiding decisions. In fact, for each moral issue in a context we can associate a landscape of potentially relevant decisions from which to learn—from the ICU on other occasions, from other ICUs, from other institutional components like IDUs, from attached ethics counselors, from other hospitals, provinces, and countries, and so on. A sound equilibrium will include an account of which of these decision sites was visited and informed the current WRE process and which were not, for whatever reasons (time, resources, relevance, and so on).

Whatever decision is made, however, a wide search for relevant, helpful information and direction will have provided evidence and argument that will have clarified, and ultimately supported the decision. The support will derive from gathering positive analogies with other ethical problem situations, modified by relevant negative analogies from these sources, all judiciously refashioned as learning continues. For example, a current situation may have as its ethical focus allocation of a scarce resource to a queue of people. Then the primary positive and negative analogies will be found in other such queue-servicing situations. For instance, the values that dominate in rules for serving queues for concert tickets and airplane boarding are different from the values that prevail for emergency exiting from an airplane and for surgery (life saving or not). Comparing these cases sensitizes assessment to which features ought to drive the ethics of a situation.

For instance, the case for continuing aggressive treatment for the young woman in the ICU might gather strength from a variety of arguments for equal opportunity to experience life (similar to priority for concert tickets) but be weak in the context of compelling arguments favoring equality of access to treatment (similar to having disabled persons board an airplane first). Any shifting of analogies, for example, from concert tickets to airline queuing, whether considered positive or negative, requires moving the ethical focus of the situation, and that must also require at least disambiguating and usually reformulating how the ethical situation is conceived, and hence what constitutes a valuable solution, and so on for all of the features involved in the cognitive problem-solving process (section 4.3).

Webs of analogies produce much stronger mutual support relationships between problems-with-their-solutions than mere logical consistency between them because the webs identify a common substantial basis for approaching web-related problems-solutions, one that has survived critical scrutiny and generated satisfactory solutions elsewhere and that has ultimately led to a valuable solution in the current case. These bonds form fundamental parts of the formation of a WRE. Indeed, the bonds reach out across ethical problem situations to form part of a pre-WRE web of mutually bonding ethical analyses that act as extended guides to WRE formation and ethical decision making, just as they do within case law and (less formally) in crime detecting procedures.

At various times a perceived unity to science, physics especially, has been the inspiration for constructing moral theory from which ethical action can be deduced; and this, ironically, despite the usually accompanying claim that ethics is very different from science. But a nice axiomatic organization to theory, assumed of science, turns out to characterize neither science nor ethics. It does not characterize science because theories are essentially idealizations—simplified versions of reality—and such idealizations can conflict, even when they are derived from the same underlying principles, because they use different idealizing principles.[18] For example, models of hydrodynamics near boundaries and in midstream free flow are incompatible, yet the two approximately converge in intermediate conditions. Molecular and physiological cellular theories are similarly related. Even the current foundational theories of physics—relativity and quantum theories—present incompatible characterizations of dynamics, the one for the very large (relativity) and the other for the very small (quantum), and are plausibly incompatible for similar reasons. The simple axiomatic conception of unity certainly does not characterize ethics either, with its extremely diverse and often mutually conflicting fundamental principles and values.

But the notion of webs of mutual support where common principles are used across situations in various subdomains, matching up at subdomain "boundaries" where the same principles analogously apply, does characterize both science and ethics and also law. So in this more complex, multidimensional, approximating sense, both may express a similar sense of unity.

There is another dimension to the kinds of unity that science and ethics manifest: in each case a variety of concepts combine to generate a richly textured collection of generalizations and applications, expressing a mutually supporting unity among them. There is, for instance, a body of such responsibilities and protocols developing around the respectful treatment of different cultures, around eco-environmental assessment, and the like. This does rule out mere collections of ideas but, important as that may be, otherwise contributes only weakly to understanding unities. In particular, it fails to illuminate the interplay between support and conflict that characterizes the ideas of science and ethics, or to tell us how to work with support and conflict. Our ethical norms are severally unavoidable, and jointly they conduce to a human flourishing that would be incomplete in the absence of any one of them; that accomplishment is ultimately the clearest way that we know they hang together. And similarly for epistemic norms. Though our norms generate endless conflicts of obligations and often require painful compromises, they combine to allow us to be whole, to support family life, public decency, mutual trust and concern, compassion, the rule of law, tolerance of diversity, and so on, all of which permits individuals to develop into mature, fulfilled persons, that is, to flourish within a genuine culture.[19] Our aim in this book is to provide a conception and tools of reason that will at least keep that vision alive and help to understand its nature.

6.4.3 Liberation, Extension, Enrichment

Having set out the structure and function of WRE, it is time to turn to its newfound power. The abandonment of foundationalism in moral philosophy liberates reflective equilibrium, first narrow and then wide, from the constraints of formal reason. Relations no longer are a matter of simple formal consistency, and support no longer is confined to deduction-from. That liberation allows reasoning to be extended—to employ nonformal reason's rich resources of observation, creative construction, multiple reasoning methods, and comprehensive critical assessment in supporting processes that develop satisfying WREs. Following from that, WRE can be extended to include an increased range and diversity of considerations that comprise robust, sensitized, analogy-based support for reflective equilibria. This allows for collective contributions from across the pre-WRE web of interrelated ethical problem solving to the understanding of appropriate resolutions of each particular ethical issue. Extending WRE to include all the resources of nonformal reason and correlatively locating problem-solving processes at the heart of WRE formation is a first major way that WRE is enriched.

That enrichment of WRE is completed with the addition of full-blooded compromise (section 6.2 and chapter 9) which expands accessible solutions to encompass all those that include one or more only partially satisfied values. While positive and negative analogy support, coupled to the resources of nonformal reason, goes some distance toward recognizing and dealing with conflict among principles and values, compromise is the key additional feature that enables WRE to take on these conflicts explicitly and with that to negotiate the rampant complexity of human life. Enriched with the resources of nonformal reason, the pre-WRE web of relationships, and the problem-solving process, compromise is liberated to devise combinations of compromised norms that represent rationally adoptable ethical positions, combinations wherein each norm may be only partially satisfied (to some partial degree and/or in some partial aspect) by the resulting action while its full obligatory force continues undiminished. For instance, chapter 9 provides a remarkable dynamic compromise that satisfies competing norms sequentially over time in a prominent public policy domain. Such compromises manifest flexible context-sensitive adjustment to complex moral dilemmas, preserve pluripotential ethical obligations that continue to orient us to our unfolding world and our life within it, and provide a broad range of options that allows boldly imaginative resolutions of problems to emerge.

Together these three features—liberation, extension, and enrichment—constitute a mature form of WRE that can play a powerful role in our private and social ethical lives. WREs become the process through which we form the contextualized equilibria that mark our ethical development as individuals and as societies, and at the same time establish the framework for reworking them when situations and lives change.

6.5 A Shift from Principles to Values

We now are in a position to address an issue that has been emerging for some time: the primacy of values over principles. The discussions of compromise and reflective equilibrium take values to be central to ethical understanding and deliberation. It is easy to see why: multiple values can define an ethical situation, whether or not they conflict; values can bear conditionally and partially on situations and on the satisfaction of other values, in ways that are multidimensionally supportive or competitive; and so on. Values can be compromised by degrees, which makes it possible to form compromises wherein some or all values are compromised to some extent. Compromises of this sort pervade our lives and are essential to understanding the ethical choices we make. Values bring much more flexibility and power to ethical decision making than do principles, which cannot be applied in such graded, multipolar ways. In addition, the shift to values is congenial to virtue ethics: virtues enter as further values to be satisfied, but with the practical rider that the consequences of their satisfaction are spread across succeeding situations. We argue that these many advantages to values together provide the best reason to adopt values as primary to formulating ethics and that the search for stronger (read: more a priori) reasons for their primacy is a chimera of formalist philosophy.

Despite these advantages, values are relatively ignored in orthodox moral philosophy.[20] Perhaps this is because the nature and functions of values cannot be captured by logical reasoning with propositions, so values cannot have the rationality of formal reason. Values cannot be assimilated to principles, yet all that is needful of ethical principles can be converted into equivalent values. Consider the ethical principle "Thou shalt not kill." What underlies it is at least the value of life. The basic obligatory or normative force of a value V is that of promoting V to the maximum extent feasible. Thus the obligatory force of the value of life is to promote life to the maximum extent feasible. While this obligation allows degrees of violation as the occasion may warrant, it also captures the force of the principled prohibition; when the latter clearly dominates, say for unprovoked killing, the value formulation carries the same prohibitive force: it is impossible to promote life when killing for no reason. Moreover, unlike principles, the obligatory force of values remains throughout their practical violations, permitting full-blooded compromises among them. And where principles fail to offer guidance, for example, in situations where someone must die no matter what action is taken, instead of a logical stalemate, satisfying the value to the maximum feasible extent would prompt an exploration of what qualities and quantities of life are involved, what wider impacts on others there might be, and so on. Values do turn up as utilities to be inserted into formal game and decision theory and optimized. But other than imposing the obligation to search for improved efficiency, the utility principle says nothing about the shape of compromises or the construction of equilibria. MEU is compatible with an undiscriminating pragmatism in which everything is always equally tradable, but it is equally compatible with the richer structure of moral deliberation developed here.[21]

6.6 Conclusion

Two themes of this chapter come together. First, there is Whitbeck's recognition of ethical problem solving as design (section 6.1). It applies to both the general character of the problem situation—how to bring about a problem resolution of sufficient value while respecting the operative constraints—and the internal problem-solving process, including the key idea of an iterative process in which problem formulation, solution values and criteria, and solution proposals all co-develop. These features exactly define how design theorists see their problem-solving process.[22] But what is an appropriate design goal of problem solving? Each solution design creates sufficient value to justify its adoption in the circumstances, but a disaggregated collection of value improvements might lose value by ineptly bringing about incoherence. However, the consideration of culture provides a conception of appropriate unity (section 6.4.2): our ethical norms are severally unavoidable, and jointly they conduce to a human flourishing that would be incomplete in the absence of any one of them. We take the development of such an ethically cultured life to be the goal of ethics as design for flourishing.

Second, there is the characterization of nonformal rationality. The discussion subsequent to section 6.1 completes our account of the two tools distinctive to nonformal reason and, with the shift to values from principles (section 6.5), that of the character of nonformal reason as a whole. To be nonformally reasonable is to be committed to applying wherever feasible the resources of observation, reasoning, creative construction, and critical assessment to the improvement of flourishing, expressed as the increasing satisfaction of value(s) and/or increasingly adequate values, aided as appropriate by the formation of compromise and mature wide reflective equilibrium.

7 Designing Deliberation

A rational judgment results from a rational process of deliberation, and rational processes of deliberation are constructed, that is, they are designed. Designing deliberation requires rational judgment itself, though, because the endeavor is rife with uncertainty, complexity, and incommensurability. Those who are deliberating want to do what is right, but in a sense of "right" much richer than validity in logic or correctness in mathematics. Judgments that emanate from deliberation have to be right for, and right to, those who are deliberating, not right for everyone. Rational use of the resources of nonformal reason is needed not only in exercising deliberation but also in designing processes of deliberation.

The best way to understand designing deliberation is to review various processes of deliberation and the use of the resources of nonformal reason in those processes. Examples from earlier chapters and new examples from social science research are instructive.

7.1 Deliberation as Dramatic Rehearsal

The women in chapter 1, at risk of having a child with a genetic condition and deciding whether to try to get pregnant, eschewed the principle of maximum expected utility and instead engaged in the kind of deliberation that John Dewey and James Tufts described in 1909:

> The notion that deliberation upon the various alternatives open to us is simply a cold-blooded setting down of various items to our advantage, and various other items to our disadvantage ... and then striking an algebraic balance, implies something that never did and never could happen. Deliberation is a process of active, suppressed, rehearsal; of imaginative dramatic performance of various deeds carrying to their appropriate issues the various tendencies which we feel stirring within us. When we see in imagination this or that change brought about, there is a direct sense of the amount and kind of worth which attaches to it, as real and

as direct, if not as strong, as if the act were really performed and its consequences really brought home to us. (Dewey and Tufts 1909: 322–323)

These women created and revised imaginative scenarios that were personal, contextual, meaningful, and challenging, and they assessed those scenarios with their intellect and their feelings in a dynamic process of dramatic rehearsal of the kind outlined by Dewey and Tufts.

Why did the women deliberate in this manner? Recall that what is important to them is the possibility, not the probability, that they could have a child with a genetic condition, and that possibility is ineliminable. They framed their problem in ways that allowed each woman to develop, refine, and probe her own distinctive situation. They imagine and reimagine what it would be like to live with a child who has a particular genetic condition, how life would be for the child and how life would change for the family and herself. Their imaginative scenarios manifest their accumulated knowledge and experience, incorporate their overt values, and engage their emotions, which express their implicit as well as their overt values. Their deliberation engages their entire being, not just their intellect. That their scenarios are highly individualistic in their themes, particular in their detail, and contextual in their development is as it should be because these women are deciding singularly for themselves, not universally for all women.

Throughout this process the women are testing their sense of their ability to accept responsibility for bringing a child with a genetic condition into the world and their ability to cope with a child with a genetic condition. Those are the issues crucial to them, and their process of deliberation is designed to confront them. They have to ask themselves many questions. How complete and realistic are my scenarios? What needs to be changed, and what is missing? What is really valuable to me? What can I compromise, and what can I sacrifice? How genuine and trustworthy are my reactions? Do my reactions change when the scenarios change, and if so, why? What kind of person do I want to be, and do I have the resources to be that person? Eventually the women have to figure out how and when they will stop deliberating. How will they know if they have a green light to try to conceive or if they are not going to get a green light? What if they cannot make a decision, and then come to feel they no longer can do anything usefully? The women need not just to deliberate but to deliberate about how to deliberate.

Dewey and Tufts allude to several features of deliberation that contribute to its rationality. One comes from the advantages of mental trials. The consequences of a mental trial are, as they put it, "retrievable," whereas the eventual "overt" trial or action, which also is a trial of the idea that produced it, "cannot be recalled" (1909: 323–324). Additionally, because many mental trials can be made in a short time, many diverse, immediate reactions—sensings, recognitions, appreciations, feelings, and emotions—can result. And, as Dewey and Tufts note, when there is a large and broad range of information, "there is clearly much greater probability that the capacity of self which is really needed and appropriate will be brought into action, and thus a truly reasonable happiness result" (1909: 324). There is no more apposite description of what those women deliberating about whether to try to get pregnant were striving to assess than "the capacity of self which is really needed and appropriate."

Another feature is the introduction of Aristotle's person of practical wisdom, the person who, as Dewey and Tufts put it, is a "good judge of what is really good" (1909: 324). The person who has become virtuous by doing virtuous acts exercises good judgment and displays the proper emotions. And the behavior of the good person becomes "the standard of right and wrong":

> There are few persons who, when in doubt regarding a difficult matter of conduct, do not think of some other person in whose goodness they believe, and endeavor to direct and clinch their own judgment by imagining how such a one would react in a similar situation—what he would find congenial and what disagreeable. Or else they imagine what that other person would think of them if he knew of their doing such and such an act. (Dewey and Tufts 1909: 324)

Dewey and Tufts do not claim that imagination, on its own, can determine what rationally ought to be done in a particular situation, but they do recognize rational dimensions of imaginative deliberation: "It helps emancipate judgment from selfish partialities, and it facilitates a freer and more flexible play of imagination in construing and appreciating the situation" (1909: 325). For our purposes, those elements promote nonformal reason's resources of creative construction and comprehensive critical assessment. They are important contributions to rationality, but what ultimately is needed is the more robust account of rationality that full nonformal reason brings to deliberation as dramatic rehearsal.

The design of the process of deliberation as dramatic rehearsal is what renders it rational, in particular, the use of the four resources (chapter 4) in the deliberation. Creative construction begins with an initial general construal of a problem and specifies that problem by imagining detailed scenarios and modifying, elaborating, and revising those scenarios in the process of figuring out what is important and assessing responses to the scenarios. Observation provides crucial, salient information for the creative construction and revision of scenarios. Formal reasoning exposes assumptions and traces the implications of possibilities that are being considered.

Nonformal reasoning methods such as induction and argument by analogy can relate hypothesized situations to past experiences and others' experiences and gauge the likelihood and desirability of their occurrence. Comprehensive critical assessment, both intellectual and emotional, scrutinizes, evaluates, and then, as judged appropriate, prompts creative construction of new or modified scenarios, only to be analyzed and evaluated in turn; and similarly prompts revision of operative values, reordering of priorities (for example, between caring and career as components of self-esteem) or introduction of new values hitherto ignored (for example, admiration of disabled people). Throughout this process well-chosen social interactions are helpful, for example, engaging with other parents of a child who might have a genetic condition, caregivers of such a child, and relevant professionals. Such experiences enlarge the range of salient observations, augment the creative construction of scenarios, enhance reasoning methods, and expand the scope and strengthen the force of comprehensive critical assessment. Normative guidance also can come from social behavior that is observed and norms that are invoked, along with requests for help from genetic counselors (although this can be frustrated by the counselors' perception of their role as nondirective information providers).

The resulting overall process is not just rational; it also evinces complete autonomy in the sense that it exhibits all four of Bruce Miller's kinds of autonomy: free action, effective deliberation, authenticity, and moral reflection (see chapter 2). The customary requirement of making an informed, voluntary decision is only part of the complex notion of autonomy; as the women's deliberations reveal, their difficulties lie in the other three dimensions of autonomy, the ones that ultimately give substance and meaning to decisions. The best these women can do (and given that "ought" implies "can," the best they ought to do) is to deliberate in the most thoroughly rational way feasible. Nonformal reason provides the capacity and the resources to do that.[1]

7.2   Deliberation as Narrative

Narratives have been introduced into bioethics as a more practically attuned, helpful way of understanding and working through moral problems. Christine Mitchell attributes their popularity to "the aridness of philosophical ethical theory, especially when applied in the analysis of actual human moral choices" (2014: S12). Much of the motivation for the turn to narratives is the desire to restore persons to the center of morality. Perhaps surprisingly, that goal represents yet another parallel between ethics and science. Paul Feyerabend, a philosopher of science, explains in the preface to his last book, *Conquest of Abundance: A Tale of Abstraction versus the Richness of Being*, why and how he tries to connect science to people:

> I avoid "systematic" analyses: the elements hang together beautifully, but the argument itself is from outer space, as it were, unless it is connected with the lives and interests of individuals or special groups. Of course, it is already so connected, otherwise it would not be understood, but the connection is concealed, which means that strictly speaking, a "systematic" analysis is a fraud. So why not avoid the fraud by using stories right away. (Feyerabend 1999: vii)

Moral philosophers, motivated by their desire to emulate what they take to be the rationality of science, likewise are preoccupied with creating and defending systematic moral theories. So the same question can be asked about ethics. Why not avoid the fraud of abstract moral systematization. Why not eschew logical or conceptual constructions and go right to stories that richly embed people in their problem contexts and link them with other people, both actually and hypothetically, thereby rendering ethics a genuinely practical human endeavor?

Doing that is the aim of those who have taken what Hilde Lindemann Nelson calls the "personal turn" in ethics (Nelson 1997: viii). The personal turn is a rejection of the powerful assumptions of logic-based analytic philosophy that ethics is universalizable, impartial, and about right conduct between strangers. Instead, personal-centered ethics has a strong particularist, not a universalist, orientation, although universal principles, often contextually hedged, can be invoked. Universalist assumptions are fostered by formal reason, and they delineate the nature and purpose of the rules that comprise morality. Lindemann Nelson readily concedes that if "doing ethics" is about developing and defending formal moral theories, doing narrative ethics is not doing ethics (Nelson 1997: xii). But if "doing ethics" means "reflecting on the moral aspects of particular encounters within a powerful social institution where what is said and done reveals a great deal about who we are and what matters in our lives," then doing narrative ethics is doing ethics, "in at least a loose sense of that word" (xii). Her philosophical caution is a perfectly understandable concession to the sway of formal reason. But her concession is unwarranted since narratives provide a richer, more robust, more productive understanding of ethical deliberation that, integrated with nonformal reason, provides a way of deliberating about, and reasoning through, moral problems in a manner focused on people, on encounters and relationships among people, and on the institutions within which people live and interact.

How is narrative ethics done? Because narrative ethics is diverse and heterogeneous, there is no simple answer to that question. We have to pick an example, and the ethnographic research that Cheryl Mattingly conducted with occupational therapists (OTs) nicely illustrates the coalescence of narratives and nonformal reason. Mattingly objects to the orthodox clinical practice of medicine, or biomedicine as she calls it, because it disregards the social and the cultural, and she traces that marginalization to a by-now-familiar source:

> It is that biomedicine's very *notion* of rationality disguises the role of human relations in constructing the clinical problem and arriving at decisions about how to treat it. Distancing from the social and moral may have reinforced biomedicine's status as a purveyor of truth, but how is this denial and disengagement possible? How can life-altering illness, something that so palpably calls the moral and existential into question, be transformed into a matter for objective rational action, for science and technology stripped of its moral, social, and cultural grounding? (Mattingly 1998: 276)

That reductionism clearly *is* possible, as Mattingly realizes, because it happens routinely in medicine. Her question is about how the clinical practice of medicine can be effective and ethical when it is no more than the application of scientific medicine. The parallel between Mattingly's criticism of biomedicine as applied scientific reasoning and our criticism of practical ethics as applied ethics is unmistakable. Both applied approaches strive to deliver the truth that only formal reason can establish, and as a result each produces a constricted conception of, and approach to, its respective endeavor. Just as there is much more to practical ethics than what is recognized by applied ethics, so, too, there is much more to clinical medicine than what is recognized by applied scientific reasoning. Mattingly's illustration of what is morally excluded from clinical medicine is striking: the "essential moral questions that plague the sufferer of serious, disabling illness—'Why me. Why now. How can I go on with my life?' And even 'Who am I now?'—are precisely the ones biomedicine sidesteps" (1998: 276). Those daunting moral questions are about how it is possible, perhaps no longer to flourish as before, but at least to live well in the midst of profound loss and adversity. As Mattingly emphasizes, as much as clinical medicine wants to evade them, these issues are at the core, not just of what medicine can do, but of the healing that medicine proclaims to be its goal.

Mattingly's alternative vision of clinical practice is exemplified by the storytelling of occupational therapists. Her ethnographic research with them allows Mattingly to examine "the way clinical practice exceeds the bounds of its own [scientific] ideology and the role clinical storytelling plays in doing so" (1998: 278) and reveals the distinct affinities that this broader conception of clinical practice has with nonformal reason. OTs believe that good treatment plans have to be devised to meet the concerns and needs of particular patients, a goal they describe as "individualizing treatment," and they regard the ability to individualize treatment as "a highly complex clinical skill" (278). Individualizing treatment is the goal of orthodox clinical practice as well, but the difference is the scope of the individualization. For OTs, exercising this clinical skill involves highly contextualized, flexible judgment that is not just medically but also socially informed and regulated, because individualizing treatment is

> a matter of judging what is best for a particular patient in a particular situation. The concern to integrate therapy within the real life of patients and their families creates a need to modify treatment in accord with singular life contexts and life histories. This "individualizing" quickly moves into the social field, drawing therapists into a complex web of social relations. Their need to tell stories is related to their need to decipher these social relations. (Mattingly 1998: 279)

Mattingly provides two examples of this skill at work.

The first concerns uncertainty about a diagnosis. A three-year-old boy has been diagnosed as having "sensory integration dysfunction," but the OT working with him suspects that he might have a head injury. The evidence for the initial diagnosis is that the child is "opposed to any kind of movement and tactilely defensive" (Mattingly 1998: 282–283). To the OT, though, he looks more like a head-injured child because "he also has a lot of things ... that sensory integration dysfunction doesn't have" (283). To try to resolve her puzzlement, the OT decides to talk to the family. She wants to hear a story about what happened to the boy. The mother relates that he was dropped on his head by an uncle, but she does not seem to be upset when she conveys that information, and she does not seem to take the incident very seriously. The mother's story and her accompanying demeanor make the OT wonder about potential neglect and abuse. The OT continues her investigation by trying to discover what happened at the time of the injury. She learns that the possibility of a head injury was not considered when the accident occurred. She also learns that the boy was brought to the emergency room, spent four hours in the hospital, and then was released with no follow-up. A pediatrician's subsequent referral to a therapist for rehabilitation mentioned only "developmental delay." The OT now constructs a story about a patient getting lost in the system. The family is poor, does not speak English, and has no insurance. Moreover, she begins to feel culpable because the boy is receiving therapy only infrequently. The family cannot afford more therapy sessions, so she cannot assess the boy's abilities properly and devise an appropriate treatment plan. She realizes that he needs to be in a school program because therapy through schools is free, but she does not know how to arrange that.

The OT is striving to understand an unclear, confusing, distressing situation by constructing stories that will make sense of what has happened, what is wrong with the boy, and how she can help the boy. But she, along with the boy and his family, is thwarted:

> The therapist found that she was lost in the story; she could not find a place in which she could be effective. The patient was lost as well. He was never properly diagnosed or treated. The family, too, was lost. They were paying for services they could not afford, and no one was helping them locate good resources for their child. (Mattingly 1998: 284)

Despite that dismaying outcome, what the OT did is rational. Mattingly regards her storytelling as part of "a reasoning process" (1998: 284). Her focus on the process is right, as is her recognition of its rationality. Given that formal reason equates rationality with reasoning, though, she is compelled to characterize the OT's storytelling as not reasoning. It is not, of course, the kind of reasoning that formal reason endorses, and it might be tempting to follow Nelson and describe it as reasoning in a "loose" sense of the word. But it is better to abandon the constraints of formal reason and appreciate the broader conception of rationality that is operating here. The OT uses the four tools of nonformal reason. She carefully observes the behavior of the boy as she is assessing him, and the demeanor of his mother as she explains how her son was injured. She uses formal and nonformal reasoning methods to make inferences about what might have happened to the boy and then critically evaluates the conclusions of those inferences, notably in interpreting the story the boy's mother tells her. Being dropped by an uncle might have been just an accident, but the mother's apparent lack of concern, in the judgment of the OT, warrants further investigation. So she entertains the possibility of culpability on the part of the hospital and devises a new process for investigating that alternative. Her creative construction leads to an understanding of how institutional, economic, and cultural factors might have contributed to the boy and his family getting lost in the system, as well as to identification of a possible source of help for the boy. In sum, nonformal reason utilizes storytelling in a rational process for addressing this problem.

Mattingly's second example illustrates how decisions about what to do can emerge from story telling. Stories can "guide future actions," in her view, because "the process of telling stories about a patient's past history or ongoing involvement in treatment is very often a process of exploring and negotiating a vision of the future good" (Mattingly 1998: 286). Telling stories to explore and negotiate "a vision of the future good" is similar to the process of constructing imaginative scenarios that was used by the women who were deciding whether to try to become pregnant after receiving genetic counseling. Mattingly's example is a collective enterprise, however, because it involves five psychiatric OTs who are trying to understand why a patient has been readmitted to their unit, in the course of which they survey alternative conceptions of the good for the patient and end up with a decision about what is best for him.

The patient is a young teenager who got upset about his mother going on a date with a new boyfriend and started clinging to her and pleading with her not to leave. The boyfriend beat him up, after which he threw a brick at the boyfriend's car and slashed the tires. Then his older brother came home and beat him up. His mother wants him put into foster care, but he wants to stay at home. Because the boy is strongly attached to his mother, the psychiatrists are worried about how a separation might affect him. The OTs point out that he does well when he is on their unit. Much happens, seamlessly and invisibly, in the one-page transcript of the conversation among the five OTs. Blame shifts as the story unfolds, and the boy clearly ends up being the victim. The OTs conclude that the boy should go to a foster home, "not because they think the boy is too much trouble, but because the family does not take care of him" (Mattingly 1998: 288). Their deliberation is brief because the story is vivid and compelling. The OTs feel no need to consider other ways of framing the problem and proceed directly to problem-solving by proposing ways in which they can help him deal with the separation.

How does so much happen so quickly and so smoothly? Mattingly's explanation is that "moral arguments are buried within clinical stories":

the therapists' interpretive process of telling the story is inextricably intertwined with identifying the morally good treatment for this patient. But the good is never explicitly discussed. Rather, it is naturalized in the very process of the telling itself, the "obvious" result of a recital of apparently relevant facts. But however invisible the reasoning process here, disguised as simply getting the facts straight, the therapists' discussion belongs much more to an Aristotelian notion of practical judgment than to clinical reasoning as applied science. (1998: 289)

Mattingly cites two crucial features of Aristotelian practical reasoning. One is that it addresses the "deepest practical question: how should one live?" and the other is that it "understands action as a *judgment* rather than an application of general rules to a particular case" (1998: 289; emphasis in original). And that judgment emanates from deliberation that is specific and contextual. It is deliberation about particular people with particular histories at a particular time in a particular place. For that reason, Mattingly explains,

> judging how to act requires capacities to "read" intentions of actors, to imagine outcomes based on a variety of possible actions, to discern what actions will be considered appropriate by others, and to understand the important historical contexts of which any given moment or any single act is just one part. Narratively speaking, practical judgment requires the actor to answer such questions as: What story or stories am I a part of here. How will these stories change if I take this act in this way rather than that one. (1998: 289–290)

Mattingly appreciates that deliberation is uncertain and risky. Indeed, the view of the OTs might not prevail because they do not have the authority and power of the psychiatrists and because the treating psychiatrist has in the past sided with the boy.

Mattingly also recognizes that "the process of story construction feels more like 'getting the facts straight' than like plotting an argument" (1998: 290). For Mattingly, explicitly "plotting an argument" is an exercise of formal reason. But instead it could be the kind of marshalling of evidence that a lawyer does when summing up in a trial or the way a detective constructs an explanation of how the crime happened. Both cases are rationally argued, certainly, and the weight of evidence is brought to bear, but both proceed from stories that cannot be reduced to simple logical inferences. Moreover, a sharp dichotomy between normative and descriptive premises cannot always be maintained. In stories, characterizations of persons can slip between describing and morally evaluating because learning can involve both, and this flexibility contributes importantly to the power of narratives. Stories and dramatic rehearsals are rationalized essentially, not by formal arguments, but by the rationality that nonformal reason accords them.

Cheryl Misak (2005, 2008) provides another illuminating account of narrative ethics, one that merges experience and narratives in ethical deliberation and is "a very practical kind of ethics" (Misak 2008: 615). Her work emerges from her experience as a patient with multiple organ failure in an intensive care unit (ICU) on full life support for a month. Like many patients in ICUs, she was in delirium. Throughout her stay in the ICU, the staff treated her as if she were autonomous and competent to consent to treatment. She believes that she should have been treated paternalistically:

"It is better to follow the policy of doing what is in the patient's best interests, despite the fact that this will be viewed as highly paternalistic by the patient herself and perhaps by much of the community of health care professionals and ethicists" (615). Advocates of evidence-based medicine and moral theorists would be stunned by her heretical view. But she learned from her experience. Misak describes narratives as "accounts that give coherence or shape to events and are thus freighted with interpretation, motivation, and other dents to what we think of as objectivity" (616). Both narratives and the experiences conveyed by narratives are rational: "Narratives ... are reason structured or open to rational criticism, and this allows them to play a legitimate and essential role in ethics," and "experience is central to rational deliberation" (618). With its focus on experience, Misak's narrative ethics is naturalistic, just as our ethics is. For her, "real life cases or real life experiences are what moral thinking turns on—it cannot do without them" (632). For her the aim of moral deliberation is "to take moral insight, judgment, and argument on their merits" (626). "If ethics is a subject on which we can legitimately deliberate," Misak realizes, "then our beliefs must be responsive to further experience, further argument, the contrary reports of others, and so on," and to do that, we need to use "the full range of our rational scrutiny" (628). There are many affinities between her conception of ethics and our conception of ethics—what ethics is and how ethics is to be done. Although Misak's view is grounded in rationality, she does not expound either reason or rationality. Nonformal reason might be a welcome complement to her moral deliberation.

Although the constructive capacity and the sensitivity of narratives have been emphasized, narratives also can powerfully shape and reinforce biased, erroneous, and ignorant (and even worse) positions.[2] Minor mistakes can be amplified to become malevolent malfeasances, heroics can be missed because they are unacknowledged, and so on. Nonformal reason is needed to scrutinize narratives critically for such outcomes, test them to limn their extent, and create alternative contexts and processes that diminish their negative impacts and enhance their positive contributions. That is, nonformal reason is used to judge the impacts of narratives, not just to judge what is in narratives.

7.3   Deliberation Shared

The conflict between respecting patient self-determination (applying the principle of autonomy) and protecting patient welfare (applying the principles of nonmaleficence and beneficence) has played out in bioethics as a debate about the locus of decision making in health care: should the responsibility for making a decision reside with the patient or the physician? Framing the conflict in that way imposes a dichotomous choice that fractures participation in both decision making about treatment and the resolution of ethical problems. Authority to decide is vested in one of the parties, with the other consequently disenfranchised. That fracture proceeds, in large part, from the presumptively sharp dichotomy between facts and values that we saw in the genetic counseling example in chapter 1. According to the prevailing roles of genetic counselor and patient, the counselor, qua professional and medical scientist, delivers the facts, and the patient provides the values. Likewise with doctor and patient. How can decision making be shared when the question is who decides, and the answer has to be one or the other. Dan Brock (1991) advises the doctor to find a "delicate balancing" between being an advocate for the health and well-being of patients and respecting their self-determination. Doctors have those responsibilities, and they have to make judgments to achieve that "delicate balancing," but it is not about decision making by doctors and patients. That decision making is not shared. The doctor decides whether to prescribe a drug. The patient decides whether the prescription is filled. Both doctor and patient can participate in discussing and examining a problem and how to handle that problem. They are sharing in a colloquy, and they are engaging in a process of deliberation that is, hopefully, well designed and collaborative. The doctor and patient may agree about what to do, but the decisions to be made are their own separate decisions.

7.3.1 Deliberating with a Doctor

In North America, the locus of decision making in medicine has recently shifted. For centuries the expertise of the doctor accorded decision making to the doctor. Physicians possess medical education, practical training, and accumulated experience, and with the emergence of professionalism, their knowledge, skills, and proficiency are tested in licensing exams and must be periodically updated to satisfy continuing education requirements. The medical expertise of physicians has been taken to entitle them to decide what is best for their patients. Recently, though, the scope of that entitlement has shrunk. Doctors, of course, still have expertise in diagnosing medical conditions and identifying appropriate treatments, but they are no longer entitled to decide whether the reasonable outcomes of treatment are worth pursuing, let alone when treatments should not be admitted because they are uncertain or controversial. Decisions about the acceptability of treatment now are determined by the values of patients. The doctor knows what can be done, and often what should be recommended, but the patient decides what will be done. Medical paternalism has largely been repudiated. Responsibility for decision making has shifted from the physician to the patient, at least in theory if not always in clinical practice. (And, recalling Misak above, not always with the patient's moral support.)

The emergence of respect for patient autonomy is represented in bioethics by models, that is, idealizations, of the physician-patient relationship. Yet the notion of sharing remains popular in discussions of physician-patient decision making. However, as long as the dichotomy between facts and values prevails and the focus remains on who decides rather than how to decide, what sharing means and how it can be achieved will remain obscure. Aspects of real sharing would be easier to identify were decision making regarded as a process that is a collaborative, mutually respectful dialogue (a) not just about information but also about how widely the situation is to be construed given the circumstances, what values are relevant to the situation (compare here the genetic counseling patients), and what alternative courses of action might be pursued and with what consequences; (b) involving elements such as negotiation and compromise; and (c) wherein the decision-making context encompasses not just making the immediate decision but projecting a cordial, productive relationship into the future. When, for example, physicians balk at prescribing antibiotics that patients request, for themselves or for a child, they might discuss not only why the patient wants the drug but why the physician is unwilling to prescribe it and mutually search for a resolution that is acceptable to both.[3] Doctor and patient can jointly participate by contributing proposals and reflectively and critically assessing them, including the values involved; by behaving in a manner that displays respect, concern, and consideration for the other person; and by accepting mutual responsibility for the position they reach.

Ezekiel and Linda Emanuel (1992) provide four models of the physician-patient relationship. The first two are familiar. One is the paternalistic model, also known as the parental or priestly model, which assigns decision making to the doctor. The other is the informative model, also called the engineering or consumer model, which ascribes decision making to the patient. The two new models are the interpretive model and the deliberative model. In the interpretive model the task of the physician is to elucidate the patient's values and help the patient choose the medical options that best realize those values. The physician functions as a counselor, examining the patient's goals, aspirations, commitments, and character with the patient in an attempt to develop and specify what is vague or inchoate and to reach a coherent overall understanding of the patient's values. The patient decides which values are important, and the physician does not judge those values. The physician then helps the patient understand what medical tests, treatments, and alternatives are consonant with those values; the patient chooses; and the physician implements the choices the patient makes based on the enhanced self-understanding that has been mutually generated. In the deliberative model, the physician helps the patient determine and choose the best "health-related" values that can be realized in a situation. The physician may discuss only "health-related" values, that is, "values that affect or are affected by the patient's disease and treatments," and may use no more than moral persuasion in promoting the health-related values the physician deems most worthy. In this model the physician is said to act as a friend.

Because each of their models incorporates a different account of autonomy—respectively patient assent, control, moral education, and moral self-understanding—Emanuel and Emanuel hold that no single model can be endorsed. They suggest that the models should be assessed, at least in part, with respect to the adequacy of the notions of autonomy they embody. Nevertheless, they claim that "descriptively and prescriptively ... the ideal physician-patient relationship is the deliberative model" (1992: 2225), adding the contextual qualification that different models may be appropriate in different clinical situations.

Both the interpretive and the deliberative models have serious problems, though. In the interpretive model, values remain the exclusive domain of the patient, and basic values are never challenged by the physician. Consequently, it is in an important sense a nondevelopmental model, because whatever improved self-understanding results is confined to clarification and not transformation—unless, as in the case of the women who, after receiving genetic counseling, determined for themselves how they would decide whether they would try to get pregnant, the enhanced self-understanding is initiated by the patient. The model posits only a weak interpretive relationship, because there is no improvement in the patient's understanding of how to enhance their suite of values or how to assign priorities to its constitutive values in varying and new contexts. A stronger model would give the *relationship* a developmental (i.e., maturity-gaining) aim by including mutual exploration of the values of the patient to support the patient and to strengthen the patient's independent critical capacities.

The deliberative model has two serious difficulties. One is the vagueness of the notion of "health-related" values. With more serious illnesses and chronic illnesses especially, virtually all of a patient's values could be affected, at least in the sense of being diminished or compromised.

Diminished or not, when profound consequences for quality of life loom, many general values, for example, relieving suffering, retaining autonomy, preserving dignity, and fostering courage, will unavoidably be in play; these reach across all the dimensions of life, including health. Given such complexity and contextuality, let alone the hazy notion of health, how can a line between health-related values and non-health-related values be drawn? Without such a line the concept of health-related values cannot circumscribe the ambit of physicians' decision making, particularly where that line is most needed. This difficulty exacerbates the other difficulty. Emanuel and Emanuel recognize that even though the deliberative physician may not go "... beyond persuasion, in practice the deliberative physician may occasionally lapse into paternalism" (1992: 2225). This attempt to design decision-making models for physicians is trapped in a dilemma of its own making: if a sharp fact/value distinction is accepted, participation in moral deliberation and decision making is fractured, and if the distinction is rejected, paternalism threatens and, moreover, bereft of the protection that health-related values are supposed to afford patients.

All four models also impose substantive obligations on doctors. In the paternalistic model doctors are obliged to put the patient's interests above their own and to seek help when their knowledge is insufficient. In the informative model physicians have the obligations to provide truthful information, to maintain their competence, and to consult with others when their knowledge or skills are insufficient. In the interpretive model physicians additionally have the obligation to take the patient through a process of self-clarification. The obligations in the deliberative model are those of a counselor, teacher, or friend. With respect to a counselor or teacher, the obligations are to inform, to clarify, and to instruct, that is, to help the patient understand and mature, but not to go beyond respectful persuasion. With respect to a friend, the obligations are to be interested, concerned, supportive, and loyal, and perhaps also include the obligations of counselor or teacher, although friendship seems to require them less centrally, and they may be in tension with being supportive and loyal, that is, with doing what is needed to preserve the friendship.

For Emanuel and Emanuel, these obligations are relative to their models, but the expansive, variegated morality of professionals and patients or clients cannot be decocted to the obligations attached to the models. Not only are all the models partial, the adoption of any model depends on higher-order value judgments that transcend the contexts of the models. In addition, value judgments are needed within the models. In the favored deliberative model, for example, determining the relevant health values in the circumstances and drawing the boundaries of respectful persuasion are matters of judgment. The role established by a model can include only selected aspects of a sweeping, open process of moral decision making, where the principal challenge remains how to properly design participation in it. Fortunately, nonformal reason can support extensive, mature developmental relationships in which locating boundaries is part of what is learned and negotiated.

7.3.2   Deliberating with a Clinical Ethicist

In his book *Troubled Voices*, Richard Zaner recounts a number of situations in which he, a philosopher trained in phenomenology and working in a hospital, provided clinical ethics consultations. At the end of the book, Zaner reflects about what these situations have in common:

> Each of them confronts the people involved—and ourselves, to the extent that we can feel the difficult issues these people faced—with a basic challenge to their sense of self. What and who we are, what we hope to be and become, even whether we will continue to be at all, is in one way or another at stake in these circumstances. This poses basic moral questions. (Zaner 1993: 137)

That richer conception of morality—a morality that is about whole persons and personal development, not just principles, about what a person can hope and strive to become, not just what are permissible and obligatory—is the morality of the women making post-counseling reproductive decisions, of K'aila's mother, and also of Dr. P. But it is not the morality of a bioethics in the guise of applied ethics.

Zaner wonders about the place of ethics in his consultations, one of which involves the Olands. The wife and mother in this family wants to have her life supports withdrawn, but the husband and father wants "everything possible" done. The matter is resolved only when, in discussions with Zaner, the husband's grief and guilt is expressed, allowing him to respond to his wife's circumstances empathetically.[4] Zaner's problem, like so many real bioethics problems, is a moral problem, but not the kind of pristine moral problem that appears in bioethics textbooks, purged of detailed circumstances and messy contexts to make way for moral theory. Clinical ethicists who equate morality with a theory can lament that one of their consultations was about a communication problem, not any moral problem. As Zaner's consult reveals, poor communication can be an important ingredient of a complex moral problem, as can lack of trust and lack of understanding. His consult does not invoke moral principles or a moral theory. Espousing autonomy is easy; helping people make difficult decisions in a coherent, sensitive, rational manner is hard. Zaner explains that the practitioner has to examine clinical encounters thoroughly "before making up one's mind about such austere matters as ethical principles" (1994: 211).

Musing on the Oland case, Zaner asks, "What did anything that went on have to do with 'ethics'—a question I've grown used to asking myself with embarrassing regularity" (1993: 32). Then he describes his job as a clinical ethicist:

> All I did was convene a meeting, moderate it, and then back out to let things take their course. I had been asked to consult by the attending physician. As he told me later, he thought that the family might feel somewhat freer to talk about things with "people like you." When he first talked with me, interestingly, he too felt quite strongly that something still unspoken was troubling this family. In asking me for a consult, he hoped that I might be able to ferret it out. He also reminded me of my own words in an earlier conversation with him when I told him that I often felt rather like a detective, and he proceeded to lay out what he took to be a "clue." When he had mentioned the possibility of a feeding tube, he had noted a kind of strained reaction. The atmosphere markedly changed, and he decided that, as this was not an immediate issue, he would be better off not pursuing the topic just then. (Zaner 1993: 33)

That "clue" does lead, albeit in an unexpected way, to a resolution of the family's troubles. But is it a moral problem, or a communication problem that perhaps would be handled better by a social worker? Zaner is brought in to deal with what looks like a prototypical bioethics problem—the withdrawal of life-sustaining treatment—but he does not wield the principles of applied bioethics. So is this one of those consultations that ends up disappointing clinical ethicists itching to do "real ethics." Zaner explains how he understands "real ethics":

> This encounter happened many years after I had begun to get involved in these sorts of situations and had already learned a good deal about them, enough to know right off that ethical issues are most often deeply buried, rarely recognized as such, and may require a good deal of indirectness to deal with them appropriately. Kierkegaard, that master of irony and the subtleties of human life, knew this best of all, for he knew enough to know that such matters can only be talked about by not talking about them directly. Indirect discourse, in his words, is the only way to go about understanding things human, especially matters of ethics. So, what I had learned to do is never to talk about "ethical issues," or rather to talk about them by never talking about them. What happened in the case of the Olands seems to fit that idea rather well: feeling that something was really bothering him [the husband and father], it seemed to me that the thing to do was just give him a chance to talk, to come to what was wrong in his own way and at his

own pace. Even when he began to talk, though, the thing lying deeply buried beneath even his terrible grief still took time to emerge, and then did so only very gradually, as if testing the light of day for any treacherous snares that might be lying about. Not talking about his grief and guilt directly was a way of letting him talk about them indirectly. People faced with such difficult decisions know this well. Such is the irony of human experience communicated. (Zaner 1993: 34–35)

Applied bioethics is deeply rooted in the logic-based tradition of analytic philosophy, the main Western alternative to the phenomenological tradition in which Zaner was trained. Irony, subtlety, and indirection are anathema to analytic philosophy, which prizes clarity, precision, rigor, and systematicity in all things. (These latter values have their proper roles; it is the framework within which they are invoked that makes all the difference.) Was Zaner wrong not to remind the family that Mrs. Oland was competent and that her autonomy ought to be respected? That would have been the open, direct, morally prescribed way to respond according to an applied bioethics conception of the Olands' situation. But does that conception of ethics appreciate the Olands' plight? Did they have a moral problem, and if so, what, other than engaging the principles of applied bioethics, would make it reasonable to regard their conflict as "moral"? Zaner found a way to reveal the momentous responses called forth by grief and guilt and then to integrate them into a healing, communicating moral integrity for the Olands. These are, as Zaner says, basic ethical issues. If he is right that ethical issues are most often deeply buried and rarely recognized as ethical issues and that the lesson of human experience is that they are best handled indirectly, applied bioethics badly misses and distorts much of morality. By Zaner's own account, his intervention was effective, but Zaner's effectiveness is not the effectiveness of applied bioethics.

What, then, is clinical ethics? How do clinical encounters go? How do conversations illuminate and elucidate? First and foremost, Zaner listens, which he calls "letting people be" (Zaner 2004: 15). Patients want to do the right thing, but their beliefs and values do not provide the clarity and direction they need. So, as Arthur Frank puts it, "they need someone to listen to them while they talk their way into new beliefs, founded on the old ones, but supplementing and revising them considerably" (Frank 2011: 157). Zaner solicits and refines stories. He is a kind, patient, sympathetic listener imbued with a philosopher's critical acumen and passion for clarity. Patients have to tell and retell stories about themselves to see where the stories go and whether the endings seem right to them. In this conversation there is a disciplined, twofold creative construction, carried out by the patients, clarifying for themselves who they are, and by Zaner, clarifying to himself the patients' situations and problems, both creations honed by sharply crafted critical appraisals. Abstract values are of little or no help here. They have to become what Frank calls "working values" (Frank 2011: 159), and values are working when they are situated in and exemplified by the particularity and contextuality of stories. The ending of a story seems right enough when it fits, that is, when it makes sense and feels right. And "sometimes," Zaner reports, "it takes a lot of telling before you get it right" (Zaner 2004: 9).

What the women who received genetic counseling do in their imagination, Zaner's patients do in their words. For both, the process is crucial to their deliberation. Doing what is right is getting the process right to get the story right. The story is the patient's story, and the judgment that the story has it right is the patient's judgment. Zaner shares in the process by helping the story to unfold and by being what Frank calls a "witness" to the telling of the story. The job of philosopher Zaner as clinical ethicist is to listen to their stories with understanding and to get his patients to tell their stories until they get them right. He is their Socratic narrative midwife, prompting them to examine their lives and using the resources of nonformal reason to render their narrative deliberation rational.

Uncertainty about what constitutes a moral problem poses a practical challenge for ethics consultants and ethics committees (and a profound challenge for the field of bioethics, given its ambition to be recognized as a profession). When an ethics consultant is appointed or an ethics committee is formed in a health care institution, procedures governing the work of the consultant or the committee have to be established. Those procedures must specify not only how work will be done but what kind of work will be done. One tricky issue, politically as well as practically, is who may have access to the consultant or the committee. An even more tricky issue is what problems may be brought to the consultant or the committee. The solution to the latter issue seems obvious: an ethics consultant or committee will consider only moral problems. But what, then, is a moral problem? Ethics consultants lament that many of their consultations end up involving communication problems rather than moral problems. They seem distressed that they are wasting their time on minor, mundane problems that should be handled by somebody else—a social worker, probably—instead of confronting the serious moral issues they were trained to resolve. How do they make that discrimination? Why is one problem merely a communications problem and another a genuine moral problem? Can distinct moral problems really be identified, isolated, and extracted, intact and whole, from the complex situations, relationships, and contexts in which they are deeply embedded? Is a clinical ethicist, as applied ethics suggests, a logician deriving the correct solution to a well defined problem, or, as Zaner suggests, a detective trying to figure out what the particular problem of particular people is? The rationality of the clinical ethicist as logician is the narrow conception of formal reason. The rationality of the clinical ethicist as detective is the broad conception of nonformal reason.

7.4 Institutionalized Deliberation

Institutional design can fracture the moral decision making of professionals. Renée Anspach's study of two intensive care nurseries illustrates that rupture between doctors and nurses. Her research proceeds from the appreciation that medical decision making is a social process, one aspect of which is how organizations and settings "shape and structure" decisions about diagnosis and treatment (Anspach 1987: 215). Anspach found that disagreements between the doctors and nurses in these units about whether to treat infants aggressively did not proceed from conflicts about the governing moral principles but conflicts about the prognoses for the infants. Even when doctors and nurses agreed about the relevant principle in a life-and-death situation, they often could not agree about the likely prognosis for an infant. This rift intrudes on ethical decision making for the infants every bit as much as does ethical disagreement.

How did these disagreements arise? From, Anspach discovered, the different kinds of knowledge on which doctors and nurses base their prognostic judgments.[5] And these different kinds of knowledge result from the different kinds of work doctors and nurses do.[6] The prognostic judgments of doctors proceed from a combination of what Anspach calls "technological cues" and "perceptual cues." Technological cues are the information obtained from diagnostic testing. Perceptual cues are the information obtained by examining the patient, using techniques such as palpation, percussion, and, most commonly, passive observation. The prognostic judgments of nurses proceed from what Anspach calls "interactive cues." Interactive cues are the information generated by social interactions between the patient and the health care practitioner. With adults, much of that information comes from a clinical history, but with infants and children, it also comes from nonverbal aspects of communication.

The types of information that doctors and nurses value and use depend on the nature of their professional training and the nature of their jobs. Doctors regularly but periodically come to the intensive care nursery to review the results of tests, read charts, examine the babies, and conduct rounds.

They spend relatively little concentrated time with the babies. Moreover, doctors change. Attending physicians have rotating shifts in the unit, and, as part of their training, residents rotate through the unit to assignments in other units. In contrast, nurses have extended, continuous assignments to patients, and they work long shifts in the unit, during which they are caring for only a few babies. They touch them, hold them, and comfort them. They see how alert they are, how much they respond to stimuli such as sounds, light, and touch, how often they cry, how "floppy" they are. Over the time they spend with their patients, they detect patterns and changes.

Anspach points out that the prognostic judgments of doctors and nurses, and the decisions based on them, emanate from "the standpoint of a very partial knowledge base, selectively filtered through the lens of structural circumstances" (1987: 227). The knowledge of the doctors is skewed toward the generalities they have been taught, learned from experience, and read in medical journals. The knowledge of the nurses is skewed toward the particularities of what they have observed in their interactions with the babies, for example, their energy levels, alertness, discomfort, and "will to live."

When compared to that of the physicians, the nurses' knowledge can be easily discounted or dismissed as idiosyncratic, biased impressions that are less relevant and reliable than the generic lessons of established medical practices. Yet the judgments of the nurses are not just subjective hunches or, in the parlance of philosophy, intuitions. One unit nurse's account of their knowledge provides a more realistic, more charitable assessment:

> I think that with a lot of nurses that have been here for a while, the gut feeling has developed from a lot of observation that they have recorded subconsciously and never put on paper, and that no one's ever written about and I think that there are a lot of those things, so when people say, "I don't like that kid, there's something wrong with him," the gut feeling's really based on a lot of informal study that they don't notice that they're doing. (Anspach 1987: 222–223)

From this perspective the views that nurses develop from their accumulated experience look much less like a "gut feeling" (or an intuition) and much more like a rational judgment that proceeds from the use of the four resources of nonformal reason. This construal of their knowledge offers the prospect of a more reasonable, more ethical response to managing conflicts between doctors and nurses.

There is, however, a further complication with the clash between what Anspach calls the physicians' perspective of detachment and the nurses' perspective of engagement, the former based on conclusions derived from diagnostic technology, physical findings, and epidemiological studies, and the latter based on inferences from extended contact and (nonverbal) social interaction with the babies. Within the culture of academic, tertiary care medicine, which is where intensive care nurseries reside, and within the history of medicine, the interactive cues perceived by the nurses and the knowledge those cues impart are, as Anspach recognizes, "devalued data" (1987: 229). The nurses' knowledge might also be devalued because most nurses are women and the knowledge of women has historically been dismissed or stereotyped as emotional, quirky, and superficial.

The fracturing of knowledge and of participation in decision making that Anspach's study reveals is a serious problem that has many ramifying institutional dimensions. Anspach appreciates that the organization of work in an intensive care nursery, within the background culture of health care, can have a deleterious impact on participation in decision making: "Those who have the most patient contact (the nurses) and the most at stake (the parents) have the least authority in life-and-death decisions" (Anspach 1987: 230). She also appreciates that the fracturing of both knowledge and participation must be removed if the quality of the decision making is to be improved:

> To the extent that decisions cannot be extricated from the social organization of the intensive care nursery, broader changes in that organization may be necessary. A culture which permits the exploration of other modes of knowledge and a social structure which facilitates greater interaction among physicians, patients, and parents, may contribute to more informed and equitable decision-making. (Anspach 1987: 230)

And, we add, could contribute decisively to genuine role accountability on everyone's part.

Although the problem of fractured participation is not exclusively a matter of institutional design, it is largely created and perpetuated by institutional design. Yet institutional design dedicated to enhanced moral decision making that goes beyond hiring a clinical ethicist and establishing an ethics committee is rare, and constitutes a daunting enterprise. Designing processes that, on the one hand, facilitate more genuinely collaborative, deliberative, and creative decision making than what prevails in the practices of ward rounds, unit rounds, teaching rounds, grand rounds, and ethics rounds, and, on the other hand, are compatible with the physician's ultimate authority and responsibility for a decision is an intimidating challenge. Likewise trying to design ways of promoting better communication and improving participation in decision making given work schedules, shift changes, and staff rotations; or trying to design the educations of health care professionals in a way that fosters more informed, complementary, and novel interdisciplinary decision making. Despite such formidable challenges, institutional design is the route to more participatory, informed, and equitable decision making. Individuals often feel alienated from the institutions with which they must deal as employees, patients, and clients, and alienation caused by features such as the unintelligibility of processes, powerlessness to affect processes, the distance between inputs and outcomes, and the lack of adaptability to local conditions has been spreading as the size, concentration, and complexity of Western institutions, both public and private, have increased. The only positive alternative to blind destruction and backward-looking rural utopian options is to improve participatory design. Properly designed participation substantially reduces the alienation of individuals by helping to reunite reason and motivation. Doing that is inherently ethical and also forms solid ground for the ethics of professional relationships, with patients and clients and among professionals.[7]

7.5 Conclusion

Matters of moment in life are matters for deliberation. To be meaningful and helpful, however, that deliberation must have an appropriate design, an appropriate overall shape and dynamics. This requirement does not impose a single, theoretical design; for example, it cannot be an abstract calculation of likely advantages and disadvantages. Rather, it must be a particular and contextual examination in which critical assessment comes not just from our intellect but from our entire being. The imaginative scenarios of dramatic rehearsal and the stories of narrative ethics are ways of deliberating relevantly and, with nonformal reason, rationally about our lives.

The preceding cases emphasize the great diversity and richness of deliberative processes. Together these processes contain the whole expression of nonformal rationality in all its internal complicatedness and its context-dependent variety, including its constant improvement. Too complicated, self-transforming, and context dependent to be reduced to any simple characterization, much less any detailed formula, deliberating is best understood through examples like those above and others found throughout the book: the women deliberating after genetic counseling in chapter 1, the vignettes in chapters 2 and 5, the ICU staff coming to terms with the encephalitis case in chapter 6, and Rapoport's debates in chapter 8.[8] This richness is no unstructured chaos; there are powerful organizing and ordering forces at work, principally the necessity that each deliberation satisfy the requirements of being nonformally rational, thus applying the four resources and utilizing compromise and wide reflective equilibrium. Moreover, the construction of a deliberation is a problem-solving activity, bringing to bear the structure of the problem-solving process (section 4.3 above) to help organize the construction of rational deliberation. This process, recall, has at its heart a learning process from which issues both particular design solutions and a general design expertise manifested in the case studies above.[9]

Deliberation is most reliable and productive when it is a collective, social enterprise. Models of the physician-patient relationship are attempts to design joint decision making, but they focus on who is entitled to be the decision maker, not on the process of deliberation. The goal of these models is not surprising because, as we saw in the presumption about how the women who received genetic counseling should make their decisions, the models likewise assume that there is a sharp dichotomy between facts and values and that rationality is reasoning from principles. Decision making can be shared only when a genuinely deliberative process is shared, as revealed in the storytelling of the occupational therapists and in Zaner's approach to ethics consultations, and then decision making is much richer and more complex than what the models of the physician-patient relationship depict. Deliberation can fail or go awry, however, when it occurs within institutional settings and constraints that do not provide the necessary opportunities and resources, and it will not occur at all when the structure or the ethos of an institution obstructs it. The only remedy for those obstacles is institutional redesign.

8   Designing Practices, Institutions, and Processes

Much of morality is socially and culturally embedded. Practices and institutions embody recognized values and ideals of the societies and cultures within which they emerge, and more profoundly they are formed and shaped by background assumptions, values, and ideals so deep and expansive that they are invisible.[1] The morality that infuses social acceptance of and participation in fundamental institutions such as school, law, health care, social media, entertainment, politics, and religious observance, and that supports diverse practices such as donating time and money to charities, respecting queues, and attending athletic events, influences how we behave, how we live, and who we become. Practices and institutions also create moral problems. Scarce health care resources raise social issues about fairness in allocation (say, of kidneys for transplantation), and personal issues about what to do when resources are not accessible (leading to responses like medical tourism for in vitro fertilization). Socially generated moral problems sometimes impose double binds that force distressing choices between two harms; the only way to escape that intractable bind is by changing or eliminating its source. Because practices and institutions pervade our lives, and because morality pervades practices and institutions, morality is in large part a matter of recognizing how moral problems emanate from the design of practices and institutions and how processes for resolving moral problems are designed. This chapter depicts those core elements of ethics as design and shows how nonformal reason is needed to account for the rationality of that design.

8.1   Fights, Games, and Debates

Anatol Rapoport's classic text *Fights, Games, and Debates* (1960), is about conflict, a topic that, according to two pioneering game theorists, R. D. Luce and H. Raiffa, preoccupies human beings more than any other, with only the exceptions of God and love.[2] Rapoport begins his examination by providing prototypes of three ways of resolving conflicts. His example of a fight is the verbal and physical combat between Tom Sawyer and a stranger he encounters as he is strolling down a street practicing a new way to whistle. His example of a game is a chess match between Frank Marshall, the United States champion, and Capablanca, the great Cuban player, in the 1918 New York Tournament. His example of a debate is the argument between Settembrini and Naphtha for the soul of a young man in Thomas Mann's novel *The Magic Mountain*. In fights, games, and debates, there are at least two participants, which can be individuals or groups—clubs, tribes, and nations fight, play games, and debate, sometimes en masse, sometimes through representatives. Fights, games, and debates represent three competitive modes of interaction with those around us. In these conflicts each participant has at least one objective, and the participants mutually interact to achieve their objectives in a manner that typically involves some combination of cooperation and opposition.

Fights are essentially oppositional, the objective being to remove the other participant from the interaction. They often proceed according to explicit or tacit rules, as, for example, in fencing duels and boxing matches, and removal may entail anything from a courteous bow to death. Games range from being wholly cooperative, through "mixed motive," to wholly oppositional, and the objective of playing the game is to maximize one's payoff (utility). Even when games are entirely oppositional, however, cooperation in following the rules of the game is required, and, because the challenge of winning the game is what makes playing the game worthwhile, the participants also must cooperate in doing their best to win. Debates are essentially oppositional, the objective being to win the approval of judges or an audience when the arguments are directed to them or to convince the opponent when the participants' arguments are directed to each other. In the terminology of game theory, fights, purely oppositional games, and debates are all zero-sum games, which means that what one participant wins is lost to all the other participants. Cooperative games are sum-increasing because by cooperating all participants increase their payoffs and hence the total payoff for the game. Fights and debates, while essentially oppositional, can be embedded in an overarching cooperative process where all participants increase their payoffs, as when a fight is staged for entertainment or a debate is used as a respectful, collaborative way to work through an issue, with primacy given to the quality of the exploration.

Before looking more closely at these three forms of conflict, it is worthwhile to note Rapoport's summary of the contest between Marshall and

Capablanca because of how compactly it illustrates the interaction of formal and nonformal reason. He begins with an analysis of what Marshall was doing years before their match, when he was becoming even more skilled by playing against himself, an endeavor of which Rapoport suggests we could aptly both "call it practicing" and "call it research":

> Until about fifteen years ago it did not occur to many people to associate research with games. Yet the thoughts of the mathematician and of the chess player are of the same sort. Both move in mazes of deduction. Stripped to essentials, the "skeleton" of these thoughts is of this nature: "If so, then so, unless so, in which case so.... But on the other hand, if so or so, then this, which leads to that ..." Put different kinds of "flesh" on these bones, and you have mathematics or logic or chess or bridge or the "elementary" reasoning of Sherlock Holmes, or the scheming of a diplomat." (Rapoport 1960: 2)

Formal deductive reasoning is the skeleton of thinking and indispensable, but it needs not just flesh but supplementation by judgment. At that time Marshall was using mazes of deduction to investigate defenses for Black to the powerful Ruy Lopez game of White, against which he hated to play. One night Marshall discovered moves that set subtle traps for White, and when he checked, these new variations were not in any of the Ruy Lopez games he could find. Marshall's quandary was when to test his discovery because if it did work in a game, it would work only once. He avoided playing the Ruy Lopez game for ten years, waiting for a proper opportunity. Capablanca finally was the deserving opponent and the worthy cooperator.

Having studied all of Marshall's games, Capablanca knew that he had not accepted a Ruy Lopez game for a long time. So when Marshall surprisingly offered the Ruy Lopez game to him, realizing that this probably was one of the most important games in Marshall's career Capablanca could surmise that he "must have investigated as much of the maze of deduction bearing on this variation as was humanly possible" (Rapoport 1960: 4). Capablanca explains his decision to accept Marshall's invitation:

> I thought for a while ... knowing that I would be subjected to a terrific attack, all the lines of which would of necessity be familiar to my adversary. The lust of battle, however, had been aroused within me. I felt that my judgment and skill were being challenged by a player who had every reason to fear both (as shown by the records of our previous encounters), but who wanted to take advantage of the element of surprise and of the fact of my being unfamiliar with a thing to which he had devoted many nights of toil. ... I considered the position then and decided that I was in honor bound, so to speak, to take the pawn and accept the challenge, as my knowledge and judgment told me that my position would be defendable. (Quoted in Rapoport 1960: 4–5)

Capablanca avoided two traps set by Marshall, then switched from defender to attacker. When it became clear that checkmate was inevitable, Marshall resigned.

Rapoport asks, "How does one reduce this drama to logical analysis?" (1960: 5). His answer is: "The game theorists have undertaken this task and in their researches have gone as far beyond what had been known about the logic of strategy as modern logicians have gone beyond the simple syllogisms of Aristotle" (5). The advances of logic and game theory are indeed impressive and important contributions to formal reason. But why must this drama be reduced to the logical analysis of formal reason? This drama is about finite, fallible human beings with feelings and values dealing with uncertainty and taking risks. Marshall had to make a judgment about when, where, and against whom to test his discovery. Capablanca had to make a judgment about whether to play an unknown variant of the Ruy Lopez game. How can those judgments, one of which led to victory and the other to defeat, be made rationally, regardless of their outcomes? Not in the logical terms of formal reason. Rather, by using the tools of nonformal reason. Poring over all those past games involves observation and comprehensive critical assessment focused on analyzing the games and testing strategies by generating mazes of deduction. Nonformal reasoning methods also are utilized, for example, in Capablanca's assessment of the dangerousness of the traps set for him and the degree to which he can rely on his history of victories over Marshall. Devising a new variation of the Ruy Lopez game is creative construction. The drama proceeds from and inheres in the judgments that had to be made. Trying to reduce it to the logical analysis of formal reason ignores those judgments and thereby destroys the drama. Nonformal reason recognizes and respects those considered judgments as rational, thereby also preserving and conveying the drama.

Participants in games and debates have to agree to an institutional structure that defines and governs their interactions. For a game to be well defined, there must be recognized participants—whether persons, companies, political parties, or others—and recognized action options for each participant; and each game-complete bundle of choices—one for each participant—must jointly lead to recognized, unique payoffs for each participant. These requirements must be stable throughout a game for utility maximization to be properly defined. And everyone involved in a game must cooperate socially to produce a stable game. Debates are similar to games, but with even tighter rules that restrict participant behavior to noncoercive, persuasive communication with turn-taking. In typical institutionalized versions of debates, communication is strictly sequenced and timed, but the idea of debate can be generalized to encompass civil discussions of issues generally, discussions extended throughout formal and informal fora across time and space. Fights, by contrast, may range all the way from ritual combat that, like the others, respects agreed rules (dueling, for instance) to warfare that even today mostly breaks any rules to secure power advantage.

For fights the primary medium of interaction is power, whether physical or psychological; for games it is utility, whether realized as a currency or as direct goods and services (economic or sociocultural); and for debates it is persuasion, whether directed toward changing beliefs, attitudes, values, principles, or behavior. The ranges of aspects open to change during an interaction are greatest in debates and least in fights. Debates permit all dimensions of a situation to be discussed, explored, and negotiated: accepted facts, values (utilities), institutional arrangements, participants, options, payoffs, and so on. For this reason debates in principle have access to the largest range of outcomes, everything from a simple win-lose to overturning the very premises of the original issue in a way that opens up hitherto uncontemplated options. They are confined only by the commitment to civil persuasion, and in the best of them—the fully rational debates—by mutual acceptance of the primacy of the quality of the joint exploration. In the parlance of chapter 7, fully rational debates are a structured form of deliberation. In practice they may be more or less intelligently pursued, but in either case they will not be rationalized by applying any narrow notion of rule-following or maximization since that cannot comprehend strategic search decisions, deep revision of assumptions and content, and the like.

By comparison, games are confined by their rules and by the maximization criteria of rationality; games cannot by definition overturn their own constitution, as debates can, or vary the objective of seeking the payoff with the maximum utility. Games must be played out as constituted, with the only room for maneuver being that allowed within their defined strategic options. Changing the constitution of a game may itself constitute a different game that is equally confined in its terms—or may be the objective of a debate or fight—but it is not playing the original game. Negotiation in a game, if any be permitted by its constitution, is confined to the game's outcome pay-offs, and its usefulness lies in the wide range of situations where maximizing utility in well-structured ways is the predominant action available. (Modern governments try to confine their financial institutions to playing well-structured, stable games that operate under agreed rules, but this goal is regularly undermined by exploitation of loopholes, ambiguities, and opaqueness, often driven by perverse incentives—in short, through poor ethical institutional design.)

Fights seem as unconfined in method as debates since, as noted, the adversaries can resort to any means useful to gaining power. However, power is a narrow medium in which to communicate, often blocking any real access even to rational self-interest, let alone to subtler questions about the presuppositions of the dispute or opportunities for reformulating the dispute. The mixture of distrust, negative emotions (often visceral), and concentration on assembling and wielding power usually ensures that rigidity. Thus the actual range of accessible outcomes is typically very narrow for fights.

Actual conflict resolution processes at whatever societal scale, from those addressed to the more limited stresses and disputes of families to those addressed to the more sweeping negotiations and wars of nations, can be, and often are, mixes of all three kinds of interaction. Dealing effectively with human interactions requires understanding their respective dynamics and possibilities and the dynamics of how issues shift among these modes. Here we wish only to make two key points.

First, institutions play key roles in characterizing and stabilizing these different modes of interaction. This much is obvious from the preceding discussion and, more importantly, makes it clear that it is only through the contributions of the embedding civil institutions that we are able to take advantage of the greater richness of games and debates. Precisely because games and debates institutionalize their interactions more thoroughly and stably than fights do, they can more readily be designed to open up new, wider possibilities for social interaction. These possibilities represent the whole weight of societal achievement that intelligence can deliver. Whether prospects of societal advancement can be attained, however, will often depend upon how much societal weight intelligence carries. In any event, without these creative possibilities for improvement we are reduced to devising intelligent methods of mutual destruction, and life is rendered poor, nasty, brutish, and short.

Institutions are as pervasive and formative in conflict resolution as they are in setting problem-solving capacities in science—where the field of cutting-edge problems is formed by institutionalized disciplinary interactions in laboratories, journals, conferences, and so on—as well as in determining the ways that problems and the terms of their resolution can be discussed and framed. These institutions and their economic host institutions, for example, universities and commercial companies, supply the funding for research and facilitate its practical application. Institutions shape problem formulation and developmental and conflict-resolution processes, constrain possibilities, and allocate both responsibilities for responding to needs (via roles) and resources to tackle problems. However, institutions are themselves fallible and improvable. They often require external diagnosis and treatment to resolve problems—for example, internal role conflicts—and that need extends from individual businesses to the governing institutions of nations or the planet—for example, the United Nations, the World Bank, and the International Monetary Fund. Institutions themselves must reflexively be the object of rational, ethical improvement as they facilitate the problem solving that itself brings improvement.

Second, richer institutionalized modes of interaction are not only more civilized, they are more ethical. Transforming a fight into a game promotes rationality over emotion and limits damage to categories that can be agreed to by the participants in the wider embedding institutions, such as the marketplace or, for politics, the whole society. These outcomes are improvements in multiple broad dimensions of ethics, for instance, procedural justice (e.g., employment benefits), individual rights (e.g., freedom of information), and the reduction of harms (e.g., pollution). In addition, institutional alternatives provide opportunities to argue for respecting various values and ethical principles as constraints on acceptable outcomes and strategies, for instance, in worker occupational health and safety—in short, to advocate for an evolution of game forms as essential to the evolution of a more ethical culture.

It is in this role that debates also shine. We can already see in debates the prospect of tackling not only the facts and rights pertinent to a dispute as they initially appear, but also the fears, mistrust, misunderstanding, egocentricity, laziness, and other divisive obstacles that underlie many disputes. Rapoport invites us to expand and enrich processes to foster situations where mutual understanding across emotional and/or cultural divides can flourish and intelligent compromise or, better, mutually transforming cooperation can emerge. To flourish, these creatively benign interactions require institutional settings that deliver mutual trust in the parties' motivation and security of action; mutually recognizable ways to communicate about values, to express appreciation of others, to own relevant personal feelings, whether helpful and not, and to communicate empathy; willingness to cooperate in the search for agreed resolutions, and so on.

In Sydney a young Muslim woman was dismissed as a day-care worker for young children over a dispute about whether she could take time to pray. Furious that her request to perform a short, simple religious duty was met initially with silence by the management, then with dismissal, she threatened a lawsuit for discrimination. Fortunately an experienced negotiator was called in. He was soon able to help each side understand how the other felt threatened and had experienced unacceptable behavior, yet despite all that retained the good will necessary to reach an agreed solution, as well as what mutual responsibilities might reasonably be reflected in solution guidelines. In the end, a simple solution was found (she stepped into a vacated quiet office for prayers while others covered child care) that satisfied everyone and cemented better communication and cooperation among staff and with management. This conflict could have played out as a legal game, likely with less harmonious and certainly with more costly results. That alternative would have been far better than resorting to fighting, but transforming the conflict into a liberating debate produced a still better ethical outcome, especially as a model for the children, one marked by mutual understanding and respect and by participatory, creative problem solving, and that introduced a paradigm for future development.

The ethical qualities displayed in this example of a debate have already been explored in some detail. In particular, the resolution of the disagreement between the nurse and physician about continuing aggressive treatment for a patient in the intensive care unit (in chapter 7) shows how the interaction of an open debate can achieve a high-quality ethical compromise. The problem there is not just about appropriate treatment, but also about how the parties to the disagreement can respect one another and take the other person's views seriously enough to warrant a compromise and to explore what would be the most effective, acceptable ethical compromise. (See also the problem for K'aila's parents, chapter 2.)

As these workplace examples show, everyone has some capacity to create institutional arrangements that enlarge the possibilities for ethical conflict resolution. That capacity is of course constrained by a person's institutional role(s): everyone can work at this within their family, but young children will have less impact than their parents; and doctors and nurses can work at modifying processes for treatment decisions, but not, in everyday circumstances, a nation's approach to health care policy. Nonetheless, to innovate as we can in institutional contexts is ethically required and the very foundation of a healthy democratic culture. In all cultures constructive contributions are tragically rarer than they could be, limited as they are by, for example, family ties or assumptions about reduced responsibility produced by racism or class, or institutions such as a club or market; but surmounting those limits vastly improves the quality—including the moral quality—of civilized life.

Not only does the progression from fights to games to debates reinforce the ubiquitous role of institutions in the constitution of societal ethics, it locates the social source of the institutional creativity that promotes ethical social life. Here is where there is room for fair-market trading to replace deceit and force in bargaining for goods and services; for hospital ethics committees to emerge as responses to tragic problems hitherto baldly faced in isolation by doctors, by nurses, and by parents; for family counseling and negotiation to precede legal disputes in a family divorce court; and so on. In short, although intelligent creative construction is a primary resource for improving ethics, it does not appear ex nihilo but must be nurtured by the kinds of institutional arrangements we have been discussing. And there are two further positives that synergistically come with such nurturing. The first is improvement in rationality; indeed, according to the account we have developed, improvement in ethics comes about through the improvement of resources for reason, especially the increase in creative constructions, such as institutional ethics committees, and richer critical assessment processes. The second is the broad array of societal benefits that flow from a successful culture of creative, constructive problem solving. The conception of knowledge-based, predominantly competitive economics that dominates our current economic phase inhibits those benefits. Creative construction, in contrast, can develop them—the managerial and policy skills required in continual evolution or revolution, for example, and the intelligent and ethical problem-solving approaches needed to manage the cultural acceptance of change—and extend them to social and political processes at large, in turn enlarging the future possibilities for further rational and ethical improvement. This process forms the natural synergy underlying adaptive resilience.[3]

8.2 Game Theory

Rapoport begins his chapter on critiques of game theory by pointing out that game theory was conceived and almost exclusively developed by mathematicians. So the nature of game theory is not surprising: "The theory can be viewed as a self-contained branch of pure mathematics—a system of theorems built up from a set of postulates … when the validity of the theorems, solutions and so on has been checked, nothing remains to be said in criticism, except possibly about the intrinsic depth or elegance of the theory" (Rapoport 1960: 226).

The alluring comfort of the necessary, certain correctness that formal reason produces is understandable. What can nonformal reason say by way of demurring? Nothing about the mathematics of a game, but something about the design and redesign of the games we strive to play as the expression of our desired society. The importance of getting the nonformal setting for effective problem solving right is wincingly brought home in Rapoport's subtle but powerful analysis of a game about trust and jealousy based on Shakespeare's *Othello*.[4] And the classic formal game of Prisoner's Dilemma demonstrates the need to restructure the problem to escape the skillfully contrived conundrum. Those two examples are considered in turn.

Innocent Desdemona represents her situation in a game in which Othello is assumed to share her moral values and has as his dominant strategy believing her to be innocent. She concludes that her best response is to affirm her innocence. This stance is morally better than falsely admitting guilt in the hope of a pardon, a response that also is of inferior expected utility so long as the probability of Othello's belief in her innocence is greater than 0.5. But Othello is suspicious of her, whether based on the vice of paranoia or the virtue-turned-vice of respecting the advice of a fellow officer whose character and motives have not been insightfully appraised. This insures that his representation of their situation is very different from hers, one in which she and he do not share the same moral values. This discordance is expressed in a game where Desdemona has a dominant deception strategy to which his tragic best response is to believe her guilty.

Nonformal judgments of similarity and dissimilarity between players can play crucial roles in negotiations and other problem-solving processes. In Othello's case, differences in nonformal judgments between Desdemona and him lead not just to differences of detail or even just to differently available strategies, but to divergent forms of the problem-solving process in which they take themselves to be participating. Whether to believe Desdemona as a matter of strategic decision making is quite different from, though subtly related to, whether to believe her as part of conceptualizing the problem itself. Here we discover a recurring theme: rationality cannot consist simply in the formal playing out of formally structured interactions, as in bargaining games and valid arguments, but fundamentally has more to do with the management of nonformal judgments about the problem-presenting situation, about oneself and others (and the relevant similarities and differences involved), and about the kinds of social interaction involved.

The classic formal game known as the Prisoner's Dilemma displays the same theme. This game and related games create situations in which individual players have conflicting incentives to compete (become individually dominant) and to cooperate (attain collective optimality), with the standard maximization solution disappointingly being the competitive one in which both players lose. For these games the truly rational response is not merely to formally analyze the game and accept the individual maximization solution, but to nonformally devise institutional or cultural circumstances that enable the game to be transformed into one where the cooperative solution becomes available at minimal collective cost. Understanding this response requires more detail about the game.

A formal game is an abstract choice situation in which each player makes a choice among available strategies and, in consequence, outcomes with positive or negative payoffs are had. In the story that gives Prisoner's Dilemma its name, two suspected criminals, P_1 and P_2, are arrested and charged with grand theft and put incommunicado in separate cells. The police offer to reward a confession with a jail sentence shorter than the maximum sentence of 10 years. If, in the face of this, the prisoners cooperate with one another and remain mutually silent, both get a sentence of 5 years on a lesser charge. If both independently deal with the police by confessing, each gets a reduced sentence of 8 years. But if only one prisoner deals with the police and confesses, that prisoner gets the minimal sentence of 3 years, while the prisoner who does not confess gets the maximum sentence of 10 years. Each prisoner's options are (a) deal with the police (Dpol) and confess or (b) cooperate with the other prisoner (Cpris) and do not confess. Since each chooses between these options independently (prisoners are kept incommunicado), there are four possible outcomes: Dpol/Dpol, Dpol/Cpris, Cpris/Dpol, Cpris/Cpris. The table below displays the abstract form of the game. The potential outcomes for each of the prisoners are in terms of years in jail, which will be represented by negative numbers to make it clear that in this case the aim of each player is to reduce their number toward zero. Then we have: $w = -10$ (penalty for unilateral cooperation with the other prisoner by not confessing while that cooperation is betrayed by the other prisoner who deals and confesses), $z = -8$ (reward, from 10 down to 8, for mutual prisoner noncooperation when both prisoners confess), $x = -5$ (reward for mutual prisoner cooperation when both prisoners do not confess), $y = -3$ (reward for unilateral noncooperation to the one prisoner who confesses and betrays the other, cooperative prisoner). In these terms, and listing outcomes in the order Prisoner 1/ Prisoner 2, the four game outcomes above become, respectively: z/z (Dpol/Dpol), w/y (Dpol/Cpris), y/w (Cpris/Dpol), x/x (Cpris/Cpris).

The gist of the dilemma is whether a prisoner should pursue self-interest at the risk of betraying the cooperation of the other prisoner. The condition that creates the dilemma is that $w < z < x < y$ (where, recall, "less than" means "worse than" because the values signify years in jail).[5] This

Table 8.1
Prisoner's Dilemma

		P_1	
		Dpol	Cpris
P_2	Dpol	z/z (–8/–8)	w/y (–10/–3)
	Cpris	y/w (–3/–10)	x/x (–5/–5)

game poses a deep paradox for rational decision: while the cooperative x/x outcome is superior to the non-cooperative z/z outcome for both players, aiming to minimize years in jail—what we might call economic rational self-interest—nonetheless drives each individual player inexorably to the worse z/z outcome. For at x/x each player has an incentive to defect from their prisoner cooperative strategy and confess in order to acquire the gains to be had (from x to y) if the other does not similarly defect, an action that worsens the other's outcome from x to w. Of course, the other player can anticipate this move and then will also have an incentive to defect in self-protection (thus moving up from w to z while moving the other player down from y to z). This reckoning moves both players to the Dpol/Dpol z/z outcome, which for each player is worse than the x/x outcome with which they began. Moreover, this prima facie irrational result seems to be forced by a plausible principle of rationality, the dominance principle: for any game player, if one strategy performs as well or better than another across all choices by other players, then it is rational to prefer it to the other. For each player in the Prisoner's Dilemma, Dpol dominates Cpris ($w < z$, $x < y$), yielding the disastrous Dpol/Dpol z/z outcome.

If this were not problem enough, there is an additional issue: z/z is also the collectively worst outcome (since $z + z < x + x$ and $z + z < y + w$). Thus the prisoners can actually do better, both individually and collectively, by choosing "irrational" non-dominant strategies.

This is essentially as far as formal rules can take us.[6] We could simply leave the matter there as another demonstration of the limits of formal reason. But we cannot, because understanding how to behave rationally in these types of situations is of huge practical and ethical importance. Situations as diverse as divorce, resource overexploitation and environmental pollution, racial/ethnic conflicts, and the nuclear arms race all display strong aspects of the Prisoner's Dilemma. We have only to contemplate the unhappy, especially ethically unhappy, non-cooperative outcomes characterizing these crossed-purposes conflict/cooperate situations to appreciate their importance for achieving quality ethical design.

It might have been this realization, combined with accepting the limitations of formal rules, that led Amartya Sen to argue that humans would be wiser to be "rational fools" by playing the cooperative strategy, setting aside the risks to reap the rewards of cooperation (Sen 1976). To follow that alternative through the wider processes of cultivating the cultures of cooperation and cross-cultural cooperation that are required to make it reasonable is to see it as one of a larger set of options for abandoning the confines of formal game theory to escape the Prisoner's Dilemma. Put so baldly, Sen's resolution lacks grounding in circumstances that could sustain cooperation against defection, thereby threatening to abandon reason in favor of arbitrary moral conviction. Although coordinated communities of absolute conviction exist, some of them inspired and peaceful, their convictions range from fantasies to fascism and include both murderous aggression and self-annihilation. Deciding which sorts of communities of conviction might provide a reasonable basis for cooperation then becomes a matter of subtle substantive nonformal judgment—a typically delicate function of the convictions concerned and their institutionalization as cultural practices. Such judgments lie at the heart of any successful society.

The more common response to the Prisoner's Dilemma is often considered formal by its advocates, but it actually derives from the pragmatic, experientially based realization that repeated plays of an interaction between players in realistic social circumstances will lead to the conditionally cooperative tit-for-tat strategy being the best for both in the long run. A tit-for-tat strategy offers cooperation as long as the other player reciprocates but punishes noncooperation with noncooperation. This approach can yield stable cooperative x/x outcomes that are regarded as rational because they maximize accumulative utility (see note 5). But the tit-for-tat strategy must deal with the ever-present temptation that individuals have to defect from the cooperative strategy, manifested overtly by deliberate deception or covertly by continual readiness to defect upon the slightest suggestion that the other player might defect.[7] How can cooperation be reinforced against that disposition?

The answer lies in those "realistic social circumstances" mentioned above. For tit-for-tat in particular to work, the communication of intentions has to be effective, and successful communication happens only in well-established institutional contexts. Reaching a cooperative equilibrium in the Cuban missile crisis (Howard 1971; see note 6 above), for example, required considerable strategic communication. Both the threat to punish should the other side not back down and a promise to respond to cooperation with cooperation had to be conveyed clearly and convincingly. In this case communication crucially shaped and correlated the mutual social expectations to establish that both sides' tit-for-tat strategies were recognized and respected. Successful communication is an important element of rational choice situations, but the judgments needed to communicate well and to play strategies such as tit-for-tat effectively are beyond the techniques of formal decision making. Judgments are made as reliable as they are by the institutionalized roles and rituals, including self-improvement rituals, from which they emanate.

Moreover, well-designed systems of institutionalized rewards and punishments can stabilize societies. For instance, appropriate rewards and punishments will convert the Prisoner's Dilemma game from a nonnegotiable game into a negotiable one where the formation of binding agreements or contracts is possible. Simply add disincentives to defect and/or incentives to cooperate until the $x + x$ outcome becomes the individually rational solution. Then individual and collective interests are reconciled. There are only two known realistic conditions under which this outcome is achieved. One is where coercion is applied in the form of an additional external penalty on defection, for example, an enforceable punishment such as a fine or jail sentence. The other condition is where coercion is imposed in the form of an additional internal penalty on defection, for example, loss of integrity, spiritual angst, and the like. Societally, the former corresponds to curbing exclusive individual interest with a rule of law backed by police and a judicial system (for example, Western liberalism), while the latter corresponds to entrenching a suitable communalist culture/religion (for example, Confucian obligation and the obligations of most other traditional societies, or communism).[8] Either way, these options show that the truly rational response does not lie in formally solving a game as presented, but in changing it through the introduction of institutional structures that provide incentives for reconciling individual and collective interests.[9]

Internal coercive arrangements must contain substantive conditions for effective communication that, in the name of the communalist culture/ religion, encourage consistency and cooperation across repeated interactions. These are the right conditions for the preferred tit-for-tat strategy to generate reliable cooperation. As noted earlier, this alternative requires substantive judgment about which institutionalized systems of conviction are reasonable, that is, possess features such as being humane and open to learning and are sustainably grounded in non-corruption. This gives them a fine ethical aspiration: societal conflict is avoided amid incorporation of persons into an uplifting culture of cooperation. Achieving this design would be a major ethical contribution. So it is all the more important that we add to our institutional design desiderata that the society be able to resiliently maintain all these ethical conditions against attempts to use cultural/religious authority to undermine them for power and self-aggrandizement. Resorting to external coercive methods and institutions is, in contrast, all too easy, all too costly, and never completely effective.

If the costs of the extremes, pure external and pure internal coercion, have generally proven large, it throws attention onto the middle-cost ground of ethical design where we attempt to retain the cultural and personal harmony of the internal alternative and the openness, innovation, and productiveness of the external alternative while designing out the instabilities of the former and the harshness and costliness of the latter. Doing this will involve designing in the right kinds of multiple, iterated interactions among the members of a community to allow them to learn clearly the social consequences of their behavior and the benefits of humanitarian care and tolerance. And so on. This approach also has its costs—primarily the requirements of imaginativeness in the creation of social processes and the constant self-organized correction of problems—but they are much more benign, including being ethically benign, and, as importantly, they permit and encourage still further exploration and learning. In practice most societies deploy a mixed solution of this general kind. We make no pretension to design expertise here, we simply want to make the point that designing such institutional arrangements is at bottom a fundamentally ethical exercise, although it is also seamlessly combined with other valuable design dimensions such as economic, scientific, and educational design.

Once the rationality of modifying or choosing a game is understood, it is easily seen to be fundamental and ubiquitous to the exercise of rational intelligence. The exercise of reason in response to the unstable coalitions of n-person games, such as those that appeared regularly in the premodern wars of competing warlords and appear now in democratic multiparty parliaments and across nations, lies in creating institutional contexts in which collectively favorable stabilities emerge. Voting for a political platform that includes the removal of incentives for tax evasion, for example, may be motivated by a desire to remove the mixed-motive game in which one has an incentive to avoid community economic responsibility through tax evasion and replace it with an incentive toward community responsibility. Similar considerations form at least a component in almost all public decisions concerning policing, economic structuring, social planning, and the like.[10] Love is a many-splendored thing, but in many cases one is forced to be a rational fool out of love only if institutional imagination fails or circumstances prevent institutional innovation.

One could hugely multiply the range of circumstances throughout life in which, rather than in the playing out of games themselves, the deeper exercise of reason and reason-based ethical design is revealed in the shaping of context—which determines the game to be played and so shapes the structure of the opportunity costs and benefits that appear—and in the choice of values, which in turn shapes all costs and benefits. Thus while choosing a strategy is context dependent because it is a response to the direct and opportunity costs and benefits appearing in each context, choosing the shape of contexts themselves is also context dependent, specifically, historically dependent, for doing so expresses the working out of a coherent life, both for individuals and for societies. This is a magnificent conjoint display of the power of (nonformal) rationality and of ethical responsibility.

8.3 Matching Problems and Processes

Lon Fuller's jurisprudential writing is replete with insightful work on institutional design. Fuller trenchantly criticized the dogged attempts of legal positivists to portray law as the product of statist power and sovereignty,[11] whether monarchical, democratic, or totalitarian,[12] and, alternatively, he strove to understand law in the broad terms of social order. Legislation, as prominent and extensive as it is, is but one legal method of securing social order. Other processes can establish rules, for example, collective bargaining, and all sorts of disputes, not just disputes about the application of rules, have to be settled to restore and maintain social order. Adjudication, mediation, and arbitration are processes for resolving disputes and conflicts. Fuller carefully examined the forms and limits of these methods, and his analyses are valuable resources for understanding ethical institutional design and the importance of selecting a method for resolving a conflict that suits the nature of the conflict.[13] Moreover, Fuller's theoretical approach to institutional design comports well with the kind of understanding provided by nonformal reason.[14]

For Fuller the design of institutions is a cooperative social endeavor based on what he calls the "natural principles underlying group life." To make this point, Fuller proposes a thought experiment: how would a group of shipwrecked people in a remote, isolated corner of the globe figure out how to live together? Fuller compounds their predicament by adding that they suffer from an amnesia that has destroyed their memory of their previous social existence. Conflicts and disputes will naturally arise among them, and, to preserve social order, someone will have to be designated to serve as a judge or arbitrator to resolve them. How would that judge or arbitrator proceed? By, Fuller answers, searching for and utilizing "natural principles":

> He would realize that it was his responsibility to see that his decisions were *right*— right for the group, right in the light of the group's purposes and the things that its members sought to achieve through common effort. Such a judge would find himself driven into an attempt to discover the natural principles underlying group life, so that his decisions might conform to them. He would properly feel that he, no less than the engineers and carpenters and cooks of the company, was faced with the task of mastering a segment of reality and of discovering and utilizing its regularities for the benefit of the group. (Fuller 1946: 378; emphasis in original)

The same approach can direct choices among competing norms. Rather than trying to determine which of precedent, custom, or equity, say, might be most legally authoritative, a judge can ask: "Which rule is best? Which rule most closely respects the facts of men's social existence and tends most to promote an effective and satisfactory life in common?" (Fuller 1946: 381). Fuller's method for addressing matters of social order is thoroughly naturalistic. His desert-island judge or arbitrator, he points out, would be stupefied to hear these "natural principles" described as a "brooding omnipresence in the skies"; rather, the principles would be regarded as comprising "a hard and earthy reality that challenged his best intellectual efforts to capture it" (379). As with the nurses in a neonatal intensive care unit for newborn babies (see section 7.4), nonformal reason provides an understanding of how those intellectual efforts can be developed, organized, and deployed and subsequently improved to become the best intellectual efforts the judge or arbitrator, in discussion with the community to be governed, can bring to the challenge of designing "an effective and satisfactory life in common" for the stranded group.

Practices and institutions are created and structured by their constitutive rules, and they operate according to their regulative rules. Games such as chess and sports such as baseball are strongly, but not exhaustively, rule-defined and rule-governed. A social practice such as gift-giving can be relatively informal, flexible, and discretionary, as in North America, or more thoroughly and rigorously rule-bound, as in Japan. Similar social and cultural variability can exist with institutions such as political campaigns and elections. Their formats, duration, and cost can be left largely unregulated and thereby amenable to strategic calculation and opportunism, or they can be strictly delimited by rules out of concerns for economics and equity.

Regardless of how extensively rule-constituted and rule-regulated an activity, practice, or institution is, judgment is central to its construction and its operation. As Fuller points out when discussing collective bargaining, "the creation of rules is a process that cannot itself be rule-bound" (Fuller 1971: 326), and that process often relies critically on experience. In the strongly rule-bound game of baseball, for example, the distance between the bases is 90 feet, and the distance between the pitching rubber and home plate is 60 feet, 6 inches. Given the nature of the game—hitting a thrown ball with a narrow round bat—and the physical capacities of human beings, in particular, their visual acuity, reaction time, and hand-eye coordination, experience has shown that those distances produce a challenging competitive game in which the best players can develop the skilled expertise of, for example, hitting a ball thrown at speeds over 90 miles per hour. The constitutive rules that specify the dimensions of a baseball diamond were not determined a priori by antecedently existing rules, but by experience. Nor are operational rules rule-bound. In chess, for example, it is a rule of thumb that a player should castle within the first ten moves of a game. That rule has many exceptions, and recognizing when to follow the rule and when to make an exception depends upon skilled judgment developed through solving chess problems, analyzing moves and games, and playing games—the combination of formal and nonformal reason used by Marshall and Capablanca.

More broadly, deciding what approach to adopt to resolve a conflict is a matter of judgment. As we saw in chapter 6, compromise is a promising method for managing a dispute when a sufficient number of the conditions favorable to a compromise are sufficiently satisfied. But when that is requires a contextual judgment. And, as we also saw, agreeing to search for a compromise is an expression of the kind of person one wants to be, which requires another set of crucial, revisable, and learnable nonformal judgments. Likewise for the methods of maintaining and restoring social order that Fuller examines. Are disputes about the custody and care of children and financial support upon the dissolution of a marriage, for example, better handled by adjudication, the traditional adversarial approach, or by some form of alternative dispute resolution such as mediation, the approach now advocated by many practitioners of family law? And what do choices about a method of resolving such personal, distressing disputes reveal about the kind of person one is and aspires to be?

Fuller's analyses of methods for resolving conflicts emphasize processes and relationships. The role of a mediator in a collective bargaining situation is, he says, "to induce the mutual trust and understanding that will enable the parties to work out their own rules." The "central quality" of mediation in general, he explains, is "its capacity to reorient the parties toward each other, not by imposing rules on them, but by helping them to achieve a new and shared perception of their relationship, a perception that will redirect their attitudes and dispositions toward one another" (Fuller 1971: 325). To accomplish that, a mediator might have to help the parties "free themselves from the encumbrance of rules" and develop "a relationship of mutual respect, trust and understanding that will enable them to meet shared contingencies without the aid of formal prescriptions laid down in advance" (325–326). This goal closely matches Rapoport's virtues of debates (section 8.1 above). As hard as shifting from rules to relationships might be for a mediator in negotiations between management and labor, it could be an even more daunting challenge for a mediator doing marriage or family counseling, yet they are tackling essentially similar design problems with essentially similar design process needs.

A relationship of trust and responsibility fostered by participation in a well designed process can have important benefits even when the immediate outcome of the process is "formal prescriptions laid down in advance," as it would be in a collective bargaining agreement. A process of negotiation conducted as Fuller envisages it, for example, can produce a common understanding that promotes harmonious governance by rules. Indeed, as Fuller relates, that mutual understanding can do even more:

> I once heard an experienced and perceptive lawyer observe, speaking of complex business agreements, "If you negotiate the contract thoroughly, explore carefully the problems that can arise in the course of its administration, work out the proper language to cover the various contingencies that may develop, you can then put the contract in a drawer and forget it." What he meant was that in the exchange that accompanied the negotiation and drafting of the contract the parties would come to understand each other's problems sufficiently so that when difficulties arose they would, as fair and reasonable men, be able to make the appropriate adjustments without referring to the contract itself. (Fuller 1971: 326–327)

The guidance and direction that rules provide is a function of what they mean, not just what they say, and their meaning and force emerges, in large part, from the kinds of background and contextual understandings that processes such as negotiation create and instill.

Despite their decided limitations and the manifold ways in which they rely on judgment and nonformal reason to be implemented, rules have a prominent, indispensable role in morality. Another virtue of Fuller's account of institutional design is that it recognizes a dynamic, reciprocal relationship between institutions and rules. The difference between a desert-island judge and a real judge, Fuller explains, is that a real judge must respect "the force of established institutions":

> If the conditions of successful group living determine the rules we ought to apply to the group, the rules already applied themselves determine in part what those conditions are. Man's nature consists partly of what he has made of himself, and natural law, therefore, demands that we must within certain limits respect established positive law. (Fuller 1946: 380)

Human beings create institutions—for example, families, schools, markets, and religions—that are compatible with existing conditions of successful group living and that influence the kinds of persons who develop under their auspices. These institutions and the kinds of persons they mold in turn modify the conditions for successful group living, thereby prompting changes to existing institutions and the creation of new institutions, and the cycling continues. Morality is a facet of this complex dynamic adaptive system, where intelligent ethical intervention consists in constructively and contextually shaping the interaction process to yield increased flourishing, not in trying to impose a timeless static outcome. Science is an equivalent epistemic facet of the process. Shi (2001) contends that public cognitive rules for science emerge from the economic roles of scientists as investors, entrepreneurs, and project managers in a research world where credibility and cash are currencies. Here, too, intelligent epistemic intervention consists in constructively and contextually shaping the interaction processes to yield increased flourishing, not in trying to impose a timeless static outcome.

One of the most pertinent aspects of Fuller's work is his recognition of the contextual applicability of methods for resolving problems. A particular problem might not be amenable to one method but be congenial to another. We will see an instance of that when we examine the allocation of kidneys from deceased donors for transplantation in chapter 9. One of Fuller's examples concerns whether the allocation of scarce irrigation waters can be handled by mediation, and another concerns whether the allocation of a collection of artwork that has been willed to two museums can be handled by adjudication.

With respect to mediation, Fuller imagines a system of irrigation canals that provides water to a hundred farmers and in which the supply of water fluctuates.[15] During an acute shortage, an equal distribution of water could result in the failure of each farmer's crops, so a decision must be made about how to allocate the available supply of water. Fuller outlines the array of considerations on which the production of a satisfactory crop depends, for example, local conditions of rainfall, heat, light, wind and atmospheric humidity, along with the balance of these factors between night and day, and the possibility of plant diseases and insect invasions. Then there are contextual factors. There might be no market for one farmer's crop because it is being abundantly produced elsewhere. And personal attributes cannot be ignored. Perhaps one farmer has a recognized expertise in coping with emergencies, whereas another tends to get drunk in times of stress. Given the plethora of relevant considerations and the complex interactions among them, Fuller concludes that the problem of allocating irrigation waters is "not even remotely suited to solution through a regime of impersonal, act-oriented rules" (1971: 335). And neither is it amenable to mediation because the interdependencies among the farmers are insufficient to provide a suitable basis for mediation, and, given the number of farmers involved, any mediation process would be overwhelmed by the number of the relationships that must be taken into consideration.

So how might this problem be managed? Fuller's solution is that the responsibility for allocating the water must be assigned to an official—a "watermaster"—who issues administrative orders that are informed and directed by consultation with the farmers whose interests are at stake. The task of the watermaster is to allocate the water "in a way that will suit the needs of the individual farmer as those needs are perceived by the farmer himself" (Fuller 1971: 336). More practically, a "conscientious" watermaster could proceed in the following manner:

> The discharge of his duties requires not simply an inspection of the farmer's fields; it requires active consultation with the farmer himself as to his needs, plans and wishes. The watermaster cannot just look at a crop and decide for himself how long it can go without water; the ability to withstand drought depends on the nature of the crop and the peculiarities of its location; these are matters likely to be better known to the farmer than to the watermaster. Furthermore, it may be desirable to give the farmer options and let him choose the one he considers most suited to his situation. The watermaster might ask the farmer, for example, whether he would rather be given today a certain amount of water or wait a week when there is a two-to-one chance he can be allotted twice as much. (Fuller 1971: 336)

The role of a watermaster, Fuller points out, differs from that of a sanitary inspector, who has rules to enforce and who can make an independent determination of whether the rules are being followed. The watermaster's task has a "mediational" dimension because the administrative orders issued by the watermaster should be "informed and directed by consultation with the man whose interests are at stake" (Fuller 1971: 336).

Fuller's examination of this problem uncannily captures the ancient Balinese role of water temples and of the head water master in regulating water among villages for the rice crop up and down hillsides, a system that scientific analyses have vindicated as essentially optimal for the circumstances.[16] Pursuing this example, which would also include examining the contrast with the modern Western commercial approach of creating a water market, is the proper foundation for understanding both the force of Fuller's considerations and the subtlety, but power, of the exercise of reason in shaping the development of dynamic complex systems over time.[17]

Fuller realized that the challenge of administering an irrigation system is ancient and universal (Fuller 1965), but he foresaw that as individual interests became "more interwoven and interdependent" in modern society, the number of problems that resemble it would drastically increase, and he specifically mentions scarce hospital resources as one example (Fuller 1971: 336). Fuller also anticipated the temptation to "judicialize" or "legalize" those problems. An advantage of doing so might be the strong egalitarianism that would be introduced. A disadvantage of doing so, however, would be the loss of a consultative, mediative approach because relying on the preferences of persons would violate the strict impartiality and impersonality of adjudication. Another advantage, though, could be its appeal to the overburdened bureaucrat, who always could resort to the defense, "Sorry, I have to follow the rules" (337).

But is the task of allocating limited resources amenable to adjudication? Fuller thinks not. He offers the real example of Mrs. Timken, a wealthy New Yorker, who, upon her death, left her collection of paintings to the Metropolitan Museum and the National Gallery "in equal shares" (Fuller 1978: 394). Her will did not specify any particular apportionment of the paintings. Fuller holds that allocating the paintings to the two museums is not manageable by adjudication because it is a "polycentric" problem,[18] a notion that he illustrates with the image of a spider web:

> A pull on one strand will distribute tensions after a complicated pattern throughout the web as a whole. Doubling the original pull will, in all likelihood, not simply double each of the resulting tensions but will rather create a different complicated pattern of tensions. This would certainly occur, for example, if the doubled pull caused one or more of the weaker strands to snap. This is a "polycentric" situation because it is "many centered"—each crossing of strands is a distinct center for distributing tensions. (Fuller 1978: 395)

What makes effecting an equal division of the paintings a polycentric problem is "the fact that the disposition of any single painting has implications for the proper disposition of every other painting. If it gets the Renoir, the Gallery may be less eager for the Cezanne but all the more eager for the Bellows, etc." (Fuller 1978: 394). Consequently, were this matter brought to adjudication, there would be no clearly defined issue to which the parties could direct their arguments. So a judge, Fuller proposes, would be tempted to adopt the classical solution: allow the older brother (the Metropolitan here) to divide the paintings into what he regards as equal shares, and then let the younger brother (the National Gallery) choose.

Recall that each gallery had a set of paintings it would like to acquire because it would complete pairings in its collection, and each preferred choice in turn may affect preferred choices among the rest of the paintings. If in that setting you allow one gallery to make all its choices before the other gallery has any chance to choose, you clearly tilt the odds in favor of the former gallery gaining many more of its preferred choices than will the other gallery. And this will remain true even if the other gallery later gets to decide which of the two sets of paintings it would choose—that requirement can never make up for the first gallery being able to freely set up a division of paintings such that, no matter which of the two sets of paintings the other gallery chooses, it would provide a better set of pairings to it than the other gallery could achieve. So the I-divide-up-then-you-choose principle does not deliver a fair outcome in this setting. Instead, as Fuller contends, you need a negotiator that will work their way across the entire collection gathering pairing preferences on both sides and attempting, choice set by choice set, to satisfy both galleries as much as possible.

Many problems have this polycentric nature. Readjusting the wage scale across a number of jobs in a factory is an example. Typically, a polycentric problem will affect many elements, whether parties or options, and involve a "somewhat fluid state of affairs" (Fuller 1978: 397). Whether a problem is polycentric is a matter of degree, of recognizing "when the polycentric elements have become so significant and predominant that the proper limits of adjudication have been reached" (398). And that, of course, makes it a matter of judgment.

Polycentric problems cannot be solved by majority rule. The owner of a factory would not allow the employees to vote on what the hourly wages for jobs should be. And game theory grossly simplifies and distorts polycentric problems. Each party can have a list of ordered preferences with the objective being to optimize the satisfaction of those preferences, but the lists do not indicate how a party's preferences might shuffle should he or she get only a percentage of one or more of the top priorities on the list.

In Fuller's view there are two suitable methods for handling polycentric problems: managerial direction and contract. But the considerations inherent in polycentric problems are so numerous and the relationships affected by such decisions are so complex that Fuller concludes, "in the practical solution of them a good deal of 'intuition' is indispensable" (1978: 398).

Individual contracts collectively solve polycentric problems when they occur within and are governed by structures that create stability and generate an equilibrium, which requires that they also create sufficient collective value and distribute it sufficiently fairly (though not necessarily equally). Obvious examples are the contracts to buy and sell made in an economic market regulated by a system of contract law. But creating a sustainable economic market and developing a coherent body of contract law are themselves formidable exercises in institutional design that require the creation of extensive, elaborate systems of rules or meta-systems processes that generate changing, context-dependent rules. Only when all this is done carefully and thoroughly do we obtain markets that improve their ethical performance.

Fuller makes two general, related points about polycentric problems that implicate nonformal reason. The first is that relying on "managerial intuition" to solve many polycentric problems does not imply that polycentric problems resist rational solutions. (Fuller 1978: 403). Fuller does not provide, or even sketch, a theory of rationality to support his claim, however. He simply suggests that rational solutions should be addressed to problems in their entirety, not to isolated parts of them. Nonformal reason and its four resources can account for the rationality of those solutions. The institution of professional baseball provides a good illustration. Managers of teams are surrounded by an array of coaches with whom they consult constantly. Managers and coaches are keen observers and monitors of the swings of their batters and the deliveries of their pitchers. Scouts for teams observe the play and assess the talent of younger players. Technologies are used to enhance observation and to analyze performance. Videotapes reveal flaws in swings; radar guns measure the speed of pitches; and computer analyses chart the location of pitches. Managers and coaches, sportswriters, and fans are avid compilers of data and calculators of statistics. New quantitative measures of performance constantly appear, and old measures are regularly narrowed and refined. Equipment is modified and improved; the roles of pitchers become more specialized; and different ways of treating the inevitable injuries are tested. Conclusions are deduced from generalizations taken to be empirically sound, and inductive inferences are drawn from records of what a player has done in similar situations in the past. Perhaps most telling of all, the failed decisions of managers are minutely examined and widely second-guessed. The institution of major league baseball utilizes all the tools of nonformal reason—observation, reasoning of all kinds, creative construction, and systematic critical assessment—in an exemplary manner. Located in that context, the "intuition" of a baseball manager does not seem so mysterious. What is called intuition looks like skilled, rational, but not infallible, judgment.

The second point endorses the observation of Michael Polanyi, from whom Fuller adopted the notion of a polycentric problem, that "it is a characteristic of living creatures that they can deal with polycentric problems, indeed, this is the nature of all 'creative' actions" (Fuller 1978: 405). Because Polanyi does not restrict his observation to human beings, he attributes the ability to solve polycentric tasks to living beings in general and animals in particular:

> On the lowest levels it may be identified with the capacity for homeostasis or purposive action, while its higher forms manifest man's power of intelligent judgment. In either case the balance is achieved by an organism reaction to the whole range of impulses that reach it from all the "centres" which it jointly takes into account. The organism evaluates their joint significance, whether reflexly or consciously, and, thus guided, produces a solution of the polycentric task, or achieves, at any rate, a measure of success in that direction. (Polanyi 1951: 176)

That this ability, as Polanyi characterizes it, grades down across simpler, evolutionarily earlier creatures is further support for a naturalistic approach to, and understanding of, nonformal reason.

Fuller's treatment of polycentric problems generalizes directly to its better-known counterpart in social decision making, wicked problems, in turn a subset of deep problems (Farrell and Hooker 2013), thereby connecting our design-for-flourishing conception of ethical problem resolution to design approaches for other kinds of problem solving and for making public policy and cognate situations. Designing a bank branch and designing a public policy for water management are social decision making problems that involve, inter alia, various interested parties with diverse values that are salient but partially mutually competing, a range of constraints that restrict available options, and a time frame for formulating and implementing a solution. Consider, for example, a proposal for stimulating commercial development that conflicts with maintaining safe, quiet streets in a neighborhood, the form of which is constrained by matters such as funding, town planning bylaws, and public health requirements, and for which there is a limited time for assessing, approving, and starting construction. In the design of both a bank branch and a water management policy, there is at least a polycentric problem in which increased pressure on some value, for example, building height for the bank branch, generates widely differing responses from other parties, such as those who object to losing sunlight or townscape to those who desire increased rental income or diner clientele next door. Coping with the polycentric nature of problems in a rational manner can be done, as Fuller shows, by using suitable methods in the context-sensitive ways demonstrated by water masters, master architects, and other proficient problem-solving professionals.

Social decision problems, however, often possess features that make them even more complex than polycentric problems. In 1973 Horst Rittel and Melvin Webber published a now-classic paper, "Dilemmas in a General Theory of Planning," in which they argue that social decision making problems often have distinctive features that make them especially difficult to solve—for example, that the conception of the solution is interdependent with the conception of the problem, that key parameters in characterizing the problem and its solution change faster than solutions can be proposed and tested in practice, and that many competing norms can be applicable to a problem makes it impossible to satisfy all of them—as well as linked in ways that invoke polycentric structure. The last excepted, these are natural extensions of the features that Whitbeck calls attention to as marking out real-time problems that must be responded to by real-time line managers (see chapter 6). Rittel and Webber conclude: "The kinds of problems that planners deal with—societal problems—are inherently different from the problems that scientists and perhaps some classes of engineers deal with. Planning problems are inherently wicked" (1973: 160).[19]

The problem-solving approach of Rittel and Webber is similar to that of Fuller in that it recognizes the necessity of judgment, endorses debates, and is consonant with nonformal reason: "[The resolution of wicked problems involves] an argumentative process in the course of which an image of the problem and of the solution emerges gradually among the participants, as a product of incessant judgment, subjected to critical argument" (Rittel and Webber 1973: 162). And they appreciate the scope of judgment: "Normally, in the pursuit of a wicked planning problem, a host of potential solutions arises; and another host is never thought up. It is then a matter of *judgment* whether one should try to enlarge the available set or not. And it is, of course, a matter of judgment which of these solutions should be pursued and implemented" (164; emphasis in original). All this suggests that approaching wicked problems within the framework and resources of non-formal reason offers the richest ways of approaching their rational resolution, if resolution is possible. In short, wicked problems are best treated as a species of deep problems; especially since it turns out that, while wicked problems present with real internal barriers to finding well-formed solutions (for example, too-rapid changes in form to follow), their core features all reduce to familiar ones involving human finitude and fallibility.[20]

8.4 Conclusion

Processes for making decisions and resolving conflicts operate within institutional settings that shape and impose constraints on those processes. Nonformal reason is needed to develop those processes, to select a process that fits the problematic situation, and to utilize the process in a context-sensitive, critically acute, productive manner. The ultimate challenge for institutional design, however, is not just getting the processes right but also getting the persons who participate in those processes right. The institutional design of processes needs to be complemented by the design of methods that foster the virtues and skills conducive to successful conflict resolution. Combined, the design of problem-solving processes and effective participation in those processes constitutes design for human flourishing.

9   Designing Policies

Public policies of all sorts substantially affect our lives and our well-being, for example, the provision of health care, education, and child care, the control of alcohol, tobacco, and firearms, the regulation of food, driving, and employment, and the availability of social services and recreational opportunities. For that reason designing policies is ethical design. The ethical import of policies of course varies—regulating the speeds and hours of truck driving is more consequential than organizing a bridge tournament— but all policies should produce goods ethically. Designing ethical policies is not simple, however, just as designing effective and efficient policies is not easy. Policy making has to satisfy an array of conflicting values—economic, legal, social, political, and cultural values, not just moral values—and the outcomes almost always are compromises. Determining which values are relevant and how important those values are in various domains and contexts requires skilled judgment. The rationality of that judgment, as we have seen, resides in how it is formed, and that depends upon how well the resources of nonformal reason (see chapter 4) are used in the process of policy making. The first example of policy making we examine, about allocating kidneys for transplantation, is an instructive instance of core problem-solving process in ethically designed policy making that utilizes the four resources of nonformal reason.

Our focus here is on the ethical design of policy for flourishing. It would be inappropriate, and impractical, to attempt any general theory of policy making. Rather, our primary intention is to illustrate the role of ethical design in policy making deliberation, and its reliance on the well-structured use of nonformal reason in framing and producing rational ethical policy. In this sense we view the impact of nonformal reason on understanding policy making as not unlike that of complex dynamics on rational policy-making process. Policies rationalized by appealing to maximum expected utility (MEU) require taking into account the consequences of actions over long periods of time, but complex dynamics can demonstrate that this can not only often be impractical, it can be impossible in principle (for example, where bifurcations arise). Such difficulties afflict climate, ecological, urban, technological, economic, and other policy formation, and thus sustainability policies generally. In consequence, it is rational to switch policy focus from accurate prediction and optimal response (MEU) to keeping options open, i.e., to maintaining or improving resilience, along with DALAL (do-a-little-and-learn) smart shorter-term management as you go. This represents a major shift of policy-making orientation. Similarly, introducing conflict-preserving compromise and enriched wide reflective equilibrium (WRE) to policy making, backed by the resources of nonformal reason, transforms the policy options available to ethical design, especially across time (section 9.1), and opens up a bold future for designed policy design processes (section 9.2).

9.1 Designing an Ethical Policy

In *Tragic Choices* (1978), Guido Calabresi and Philip Bobbitt present an elegant study of how three societies—the United States, England, and Italy—manage tragic conflicts between fundamental values. Because fundamental values cannot be abandoned, when they conflict they cannot all be fully satisfied. Conflicts between fundamental values force tragic choices about how those values are to be partially satisfied, and the failure to fully satisfy those fundamental values leaves unresolved moral residues. Neither the situations that impose tragic choices nor the responses to those situations are themselves immoral, because complete, final resolutions of conflicts between fundamental values are not possible. Calabresi and Bobbitt recognize the inherent complexity, contextuality, and fluidity of making policies that require tragic choices:

> The object of public policy must be ... to define, with respect to each particular tragic choice, that combination of approaches which most limits tragedy and which deals with that irreducible minimum in the least offensive way. Of course, that combination will vary, not only over time ... but also from society to society, since the object is to find the approach which is least destructive of values fundamentally held in each society. (Calabresi and Bobbitt 1978: 149)

The complexity of tragic choices requires judgments about which decision-making approaches are most apt and how those approaches can be most constructively melded. The contextuality of tragic choices requires judgments about which particular fundamental values are relevant to a particular tragic choice in a particular society at a particular time and how a society's array of fundamental values can be best preserved and, when necessary, breached, but with the least possible violation.

Even though the tragic choices of both individuals and societies seem beyond reason, they need to be rational, not arbitrary or irresponsible. Sophie's choice is the memorable fictional example (Styron 1979/1999). An SS doctor at Auschwitz forces Sophie to choose which of her two children will die by threatening to send both to the gas chamber if she does not choose. Jillian's choice was real. Jillian Searle is an Australian mother who was caught with her five-year-old son, Lachie, and her 20-month-old son, Blake, in the tsunamis that swept over Phuket in 2004. In a television interview she said she knew all of them would die if they stayed together, so she "had to let go of one of them and I just thought, I had better let go of the one that is oldest" (Blatchford 2005: A3).

As Ms. Searle's explanation reveals, her tragic choice was not beyond reason. In the interview she discloses that Lachie cannot swim and that he "is petrified of water—even the pool at home." (Blatchford 2005: A3) Those features of their desperate situation compound the tragic nature of her choice but do not move it beyond reason. That a five-year-old child has more capacities than a 20-month-old infant, and thus a better chance of surviving, remains a legitimate reason for letting go of her older son. Miraculously, Lachie was able to cling to a pole or doorknob until a second tsunami receded, after which he was found by a security guard. Sophie's dilemma is worse. She, too, can only hope that a miracle saves the child she chooses, and there is no morally tolerable reason for choosing between her son and her daughter. Sophie is tortured in an exquisitely nasty and morally despicable manner.

Policy makers are removed from the devastating heartbreak, sorrow, and guilt of personal tragic choices. Their tragic choices on behalf of society enact generic policies or perpetuate the status quo, both of which create faceless vulnerabilities and distribute anonymous risks. Drafting soldiers in wartime, for example, exposes unknown persons to the risk of an early death, and failing to provide adequate nutrition and prenatal care for pregnant teenagers jeopardizes the well-being and lives of unknown women and their unknown children. How can societies propagate policies that affect their citizens so extensively and so irrevocably and implement those policies without resorting to extreme coercion? To be accepted, the policies have to be, and be regarded to be, absolutely necessary, effective, fair, and responsible. There are, however, various notions of effectiveness and multiple forms of fairness, none of which can dominate the others to prevail unstintingly, and which inevitably conflict. Consequently, the moral burden of legitimacy falls not just on the tragic choice itself but on how the tragic choice is made. The rationality of tragic policy making is the rationality of nonformal reason, which, as we know, is the rationality of the processes from which choices emanate.

The seminal work of Calabresi and Bobbitt examines the diverse ways in which the United States, England, and Italy devised policies about conscription, limits on childbearing, and, before the advent of transplantation, the allocation of kidney dialysis machines. They focus on the fundamental values of a society that tragic choices expose—for example, the preciousness of life, equality, efficiency, and honesty—and the various methods that can be used to make tragic choices, for example, a lottery, a market, a responsible governmental body, and a nonresponsible "parajury." Making a tragic choice requires ethics as design—designing an approach that, in the prevailing circumstances, affirms the fundamental values that need to be affirmed and degrades the fundamental values that have to be degraded in the least destructive manner. Accomplishing that, as we will see, requires compromise for the same reason we saw in chapter 6—none of the conflicting values can be abandoned, yet a choice has to be made. The compromises in designing policies for tragic choices are twofold. Compromise occurs in a process of developing an approach for making a tragic choice *at a time*, and because that compromise leaves a moral residue, compromise recurs throughout a sequence of approaches for making that tragic choice *over time*.

Our examination of policies that make tragic choices begins by clarifying the notion of a tragic choice. How does a tragic choice differ from a tough choice that could bring great harm, for example, a military officer's decision about whether to hazard an attack on an enemy position or a patient's decision about whether to try a risky, experimental treatment for an illness? And what implications does the tragic nature of a choice have for how that choice should be made? For a compelling practical example of a tragic choice, we review the allocation of kidneys from deceased donors for transplantation in the United States, where a long-established policy has recently been replaced. To understand and assess the change, this tragic choice is located within the work of Calabresi and Bobbitt, fittingly, because this issue is the successor to their study of the allocation of kidney dialysis machines.

Calabresi and Bobbitt conclude that the three countries they examined retain fidelity to their distinctive fundamental values by adopting an approach that is remarkable because it would not even be considered in formal reason's prescribed application-of-principles method: cycling through their fundamental values, satisfying now one, then another, while emphasizing that each is an expression of a fair and highly beneficial societal choice. Moral compromise as developed and defended in chapter 6 can be used to develop such a policy for allocating kidneys. The dual operation of compromise in an approach to allocating kidneys at a time, and in cycling through approaches to allocating kidneys over time, can limit tragedy by degrading society's fundamental values in the least destructive manner.

9.1.1 The Nature of Tragic Choices

Distressing choices are, as Calabresi and Bobbitt recognize, sometimes tragic and sometimes not tragic. What is the difference? Why are tragic choices not just hard choices about big stakes? Calabresi and Bobbitt do not offer any "simple definition to separate the difficult choice from the tragic, or the trivial from the difficult"; instead, they provide examples of choices they take to be tragic in particular societies and rely on the ability of readers to identify tragic choices in their own societies. In introducing their comparison of how the United States, England, and Italy allocate renal dialysis units, for instance, they describe the problem as "a paradigmatic tragic choice" (Calabresi and Bobbitt 1978: 177). The tragedy clearly manifests the existence and the character of the residues left by partially unsatisfied values, which can assume many different forms and possess various intensities, as the introductory examples of the military decision and the patient's treatment decision illustrate, and can only be judged in their contexts.

But even if a tragic choice cannot be precisely and exhaustively defined, it can be elucidated. Distributions of scarce goods that entail "great suffering or death," Calabresi and Bobbitt explain, "arouse emotions of compassion, outrage, and terror," which reveal conflicts between, on the one hand, the source of the scarcity and the values used to determine the recipients of the scarce good and, on the other hand, "those humanistic moral values which prize life and well-being" (1978: 18). It is hard to imagine suffering worse than the great suffering of living through the latter stages of amyotrophic lateral sclerosis (Lou Gehrig's disease), for example, but the suffering that attends a life on dialysis and shadowed by death, if not comparably great, nevertheless is vast. Still, we all die, but not all deaths are tragic, and even great suffering might not be tragic if it somehow can be redeemed.

Moreover, great suffering and death might not be the outcomes of tragic choices. Tragic choices can be tragic when their outcomes are not tragic, as the different outcomes of Sophie's choice and Jillian Searle's choice demonstrate. One of Sophie's children died as a result of her tragic choice. The loss of her daughter is calamitous, grievous, and desolating for Sophie, but that tragic outcome is not itself what makes her choice tragic. Both of Ms. Searle's children survived despite her choice, yet her choice is tragic. The nature and source of a choice insofar as they affect the chooser(s), not just the outcome (or, for Ms. Searle, not at all the outcome), are what make a choice tragic.

The tragic nature of Sophie's and Ms. Searle's choices stems from conflicting fundamental values. Like Sophie's choice, Ms. Searle's choice pitted her love for one child against her love for another child. Whatever choice either made would violate the foundational ideals and obligations of motherhood. Good mothers, because they are mothers, care for and treat their children equally. Both Sophie and Ms. Searle were forced to subjugate their love and their loving for one child to preserve their love and their loving for another child. They had to repudiate constitutive values of motherhood that cannot be repudiated. The tragedy of their choices lies in the irresistible acceptance of the unacceptable—that even consummate maternal love cannot remain inviolate in the face of fate. Such tragic choices leave a residue because they do not resolve but perpetuate the conflict that necessitated the choice. Sophie's love for and commitments to Eva do not die with her daughter, and Sophie's despair about her breach of her maternal responsibilities will never disappear. Styron's novel recounts the subsequent moral and personal disintegration of Sophie. Ms. Searle will not be similarly devastated, but she will forever carry the tragic choice to abandon her son.[1] Her regret and guilt can be assuaged but not expunged.

The source of a tragic choice also can be part of its tragic nature. The tragic choices of Sophie and Ms. Searle are externally imposed, Sophie's by society and Ms. Searle's by nature. Tragic choices also can be self-imposed. In the afterword to an anniversary edition of *Sophie's Choice*, William Styron explains that the book was partly inspired by *Five Chimneys*, a memoir written by Olga Lengyel, who was transported to Auschwitz with her family in 1944. Of the horrors that Ms. Lengyel graphically and starkly describes, the one Styron finds "most chilling of all, somehow, surpassing the butcheries and beatings," is

> the description of the author's arrival at the camp in a boxcar, and the decision she was forced to make about her mother and one of her children. Confused, and unaware of the lethal workings of the selection process, Lengyel lies about her twelve-year-old son's age, telling the SS doctor that the boy is younger than he is, in the mistaken belief that this will save him from arduous labor. Instead of being spared, the boy is sent to the gas chambers, along with his grandmother, whom

Lengyel, again in ghastly error, helps kill. She asks the doctor that her mother be
allowed to accompany the child in order to take care of him. (Styron 1999/1976) [2]

In the horrors of the situation, her confusion, and her ignorance, Ms.
Lengyel makes two tragic mistakes. In almost any other circumstances, her
lie about her son and her request for her mother would be understandable
and reasonable. She is a loving mother and a loving daughter, who, caught
in desperate straits, desperately wants to care for and protect her son and
her mother. But in a foreign, depraved world of relentless barbarity and cru-
elty, she cannot both save them from misery and save their lives. She will
be consumed not just by grief and remorse for the momentous losses but by
self-blame and guilt for the misunderstandings that brought those losses.
She will hold herself culpable. Her flaws are not the flaw of hubris that
inevitably dooms the protagonists in Greek tragedies, but they are tragic
flaws nevertheless.[3] The outcomes of her choices were truly tragic, but so
were her choices.

The stakes in tragic choices are not just high stakes but high moral
stakes for the chooser(s). For Calabresi and Bobbitt tragic choices implicate "humanistic moral values which prize life and well-being" (1978: 18).
The tragic choices of Sophie and Ms. Searle involve conflicts between precisely those momentous values, none of which can be personally or morally
renounced. Calabresi and Bobbitt provide a criterion for a satisfactory resolution of a tragic moral conflict: that it limit tragedy as much as possible by
minimizing the violation of fundamental moral values in the least offensive way. Calabresi and Bobbitt do not explain what "the least offensive
way" means, but it is plausible to surmise that properly managing the violation of fundamental moral values has both a substantive and a procedural
dimension. The substantive constraint prohibits outcomes that "implicate
moral contradictions" and are "morally debasing," and outcomes that are
arbitrary or discriminatory (18). The procedural requirement mandates a
process that fits the nature and context of the particular tragic choice that
has to be made, but only when, of course, the situation allows. The crises of
Sophie, Ms. Searle, and Ms. Lengyel did not allow that, but most policies are
not made in crises, though some are made in self-created crises.

Because none of the fundamental values that conflict can be abandoned,
a tragic choice must somehow retain fidelity to all the values involved. For
the inescapable, singular tragic personal choices of Sophie, Jillian Searle,
and Olga Lengyel, fidelity to foregone motherhood is expressed by unending regret, accompanied by varying forms of remorse and guilt. Those emotional manifestations of the moral residues of their choices are enduring

affirmations of the enormous value of what they had to sacrifice. For Sophie and Olga Lengyel there is no salvation; for Jillian Searle there is solace.

Making tragic social choices in the least offensive way requires that those choices emanate from a rational decision-making process and that the operation of compromise in that rational process retain fidelity to all of the conflicting fundamental values. In policy making at a time, a compromise minimizes the tragic outcome of not being able to satisfy all of the conflicting fundamental values by satisfying each proportionately to its importance in the current circumstances. In policy making over time, compromise takes the form of cycling through the conflicting fundamental values by giving each its turn at primacy. With rational processes for constructing such compromises, tragic choices about allocating kidneys for transplantation are not beyond reason.

Because the tragic moral choices of Sophie, Jillian Searle, and Olga Lengyel concern intimate personal relationships in all their dimensions, their resolutions will be distinctively personal. Tragic choices about policies for allocating kidneys for transplantation, like all public policy choices, concern limited generic features of and relations among candidates—for example, age and time in a queue—and in those respects are impersonal. For our purposes the upshot is that both personal and policy choices use compromise to make tragic choices, thus we see them as differing only in degree, not in kind. We leave any debate about whether there are further morally (as opposed to simply psychologically) salient differences between them for another occasion.

9.1.2 Making Tragic Choices

Kidney transplantation extends the lives of recipients and vastly improves their quality of life, thereby engaging what Calabresi and Bobbitt call "the principal humanistic value at stake … in every tragic situation, life, or its correlative, well-being" (1978: 23). Because kidneys from deceased (and living) donors are scarce, kidney transplantation also engages the fundamental values of equality and efficiency. Societal choices to restrict procurement of kidneys to those from deceased donors who had signed a donor card (opting in rather than opting out) and to allow family members to override that permission are choices that substantially create the scarcity of a resource that alleviates suffering and postpones death. In doing so the society creates subsequent tragic choices for itself: the ensuing societal choices about how kidneys from deceased donors will be allocated are tragic choices that determine whose suffering will be abated and whose life will be extended.

Calabresi and Bobbitt distinguish those two independent but related tragic societal choices. How much of a scarce good there will be is a "first-order determination"; who will get the scarce good is a "second-order determination." The scarcity of kidneys for transplantation results from a panoply of first-order social determinations that have been made about the medical, institutional, and financial resources that are available for organ procurement and transplantation and the policies that control and limit the donation of kidneys. They also have been made by the culturally embedded eating, drinking, and exercise patterns of that society, and many other factors that contribute to the ultimate level of scarcity that emerges, but here we focus on the health care roles. Those choices manifest a diversity of fundamental values. Prominent among those values is individual free choice, which is the basis of an opting-in procurement system, but also prevails in the offer-and-acceptance practice of allocating kidneys. When a kidney becomes available, it is offered to the eligible candidate on the waiting list, who then has the option of accepting it or declining it and waiting for a higher-quality kidney. Which candidate ultimately gets to make that choice is the result of second-order policy determinations of (a) which of various methods will be used to allocate kidneys and (b) how those methods will be combined.

Calabresi and Bobbitt scrutinize the principal methods for making tragic choices and ways in which these methods can be modified: a pure market and a modified wealth-distribution-neutral market; a pure lottery; a pure responsible political agency; and an "a-responsible" agency or jury. Any form of a market is precluded in kidney transplantation as long as the buying and selling of kidneys is prohibited. A lottery is possible, but, as Calabresi and Bobbitt observe, it is "a choice not to choose" (1978: 41). A procedure and guidelines for allocating a scarce good can be established through an existing political process and can be responsible if a justification for how much of the good is being provided and who is getting the good was expected and is actually provided. An "a-responsible" agency or jury is representative and decentralized, and does not give reasons for its decisions (57). Like juries in a legal system, "parajuries" are groups of individuals deemed representative of the community who, in private, make a collective decision without having to provide any explanation of, reasons for, or defense of that decision. The "parajury" has been tried: before the development of kidney transplantation, the so-called Seattle God Committee decided who would receive the scarce resource of access to a kidney dialysis machine.

Procedural and substantive values infuse both the methods of making tragic choices and the approaches to deciding what methods should be used and how those methods should be combined. Different methods, as Calabresi and Bobbitt recognize, have different strengths and different weaknesses. A pure lottery, for example, is objective, certain, simple, clear, and completely egalitarian. A responsible political agency is centralized and distanced, whereas an a-responsible agency or parajury may be decentralized, hence more attuned to the values of a local region. Moreover, different procedures can impose different substantive costs. A method that seems arbitrary and capricious because it does not explain decisions creates anxiety and frustration. Uncertainty about how and why decisions are made also can compel claimants to reveal intimate, hopefully persuasive, details about their lives. Calabresi and Bobbitt describe that burdensome, demeaning imposition on anxious, vulnerable people:

> The more crucial the result, the more pressure one feels to bare one's soul to try to gain a favorable judgment; and the more one can perceive real or imagined differences in the relative ability to state one's case, the higher these frustration costs are likely to be. And these are borne by the very people who already bear the costs of losing out on the allocation of the tragically scarce resource. (1978: 61)

The values displayed in the process of designing methods into an approach for allocating kidneys—openness, transparency, and honesty—contribute strongly to fostering and sustaining the public trust that such a tragic choice vitally requires.

Ultimately, however, the respective importance of the substantive values of equality and efficiency in allocating kidneys must be settled. Tragedy is to be limited as much as possible, but tragedy can only be limited, not eliminated, so the remaining irreducible minimum of tragedy must be dealt with in the least offensive manner. Tragic choices about the forms of equality that are relevant to allocating kidneys and the comparative importance assigned to equality and efficiency are guided by the goal of impairing the fundamental values of society as little as possible.

But tragic choices are never stable. Calabresi and Bobbitt quote a literary critic's view of tragedy: "Basic to the tragic form is its recognition of the inevitability of paradox, of unresolved tensions and ambiguities, of opposites in precarious balance. Like the arch, tragedy never rests."[4] Fundamental values that have been displaced or only partially satisfied will eventually be reclaimed, and fundamental values that have been prized cannot long escape being relegated. So how is the dynamic nature of tragedy handled?

9.1.3 Approaches to Allocating Kidneys for Transplantation

On October 23, 2012, in the United States, there were 116,186 candidates on the waiting list for a kidney transplant, of which 74,126 were active candidates.[5] Active candidates are eligible to be considered for organ offers; inactive candidates are either medically unsuitable for transplantation or need to complete other eligibility requirements. From January to July, 2012, there were 8,280 donors, and 16,586 transplants were performed. Allocating organs from deceased donors for transplantation is the process of determining which patients will be put on the waiting list for a transplant and the sequence in which kidneys that become available will be offered to candidates on the waiting list. Different organ allocation systems have different goals.

When kidney transplantation was being introduced and its success was still unproven, the obstacle that had to be overcome was the body's rejection of a foreign kidney. To reduce the risk of having transplanted kidneys rejected, tissue-type testing was used to assess how closely a candidate "matched" a donated kidney, and matching continues to be used in allocation decisions. With the development of anti-rejection drugs, however, the importance of matching decreased, and considerations of fairness became relevant. In the United States kidneys have long been allocated primarily on the basis of first come, first served, a familiar and simple way of according equal respect to persons in the distribution of scarce valuable goods. Recently the Kidney Transplantation Committee of the Organ Procurement and Transplantation Network (OPTN), on behalf of the United Network for Organ Sharing (UNOS), has developed recommendations for improving the allocation of kidneys and has adopted a new system for allocating kidneys.

The Previous System The previous system for allocating kidneys assigned points to rank-order candidates on the waiting list. Everyone cumulatively received one point for each year on the waiting list. Candidates who are highly sensitized, i.e., who are substantially less likely to be matched with a kidney or who are substantially more likely to reject a kidney, received additional points. Because it could take a long time to find a suitable donor for highly sensitized candidates, they were assigned four points. The process of matching kidneys and candidates to predict the likelihood of kidney survival compares the human lymphocyte antigens of donor and candidate (HLA matching). Each person has three pairs of these antigens—A, B, and DR—so the maximum match or mismatch is six. The fewer the mismatches, the greater the likelihood of kidney survival, so having zero mismatches is best. The extent to which the A, B, and DR antigens affect kidney survival varies, however, with the DR antigen having the strongest influence. Candidates received one extra point for having a 1 DR HLA mismatch and two extra points for having a 0 DR HLA mismatch. Relatively few candidates got two points; more candidates got one point. Candidates who had been living donors were rewarded with four extra points, but they are rare. Overall, time on the waiting list was the major determinant of priority for a transplant.

This approach had two perceived weaknesses that prompted proposals for improving it. One is that it did not seek to optimize the length of kidney and candidate survival by matching the donated kidney's likely longevity and the recipient's likely longevity. A kidney capable of functioning for decades could, for example, have been allocated to an elderly candidate who had only a few years to live. The other weakness is that it did not minimize death on the waiting list because it did not recognize that different candidates have different prospects of surviving the wait for a transplant.

The previous system did not promote the value of efficiency—getting the most post-transplant life-years for kidneys and recipients from a scarce resource—and it did not recognize the value of rescuing lives. The value of equality, in the temporal form of first come, first served, dominated the system.

The Original 20/80 Proposal An initial proposal for revising the previous allocation system combined two different methods for matching kidneys and candidates—survival matching and age matching—that are designed to increase the longevity of both kidneys and recipients. A kidney donor profile index was developed to assess the quality of a donated kidney and, on that basis, to estimate how long the kidney is likely to function post-transplant. An estimate of the length of post-transplant survival of candidates also was created, based upon four factors: age, length of time on dialysis, any prior organ transplant, and diabetes status. The highest-quality kidneys presently comprise 20% of the pool of donated kidneys. Survival matching offers the 20% of kidneys with the longest expected length of function to the 20% of candidates who have the longest expected length of life. Age matching is used to allocate the remaining 80% of kidneys, which are offered to candidates who are between 15 years older and 15 years younger than the donor. Whereas equality is the dominant value in the previous allocation system, the combination of survival matching and age matching makes efficiency—correlating the projected longevities of kidneys and candidates—the dominant value in the original 20/80 proposal.

The Amended 20/80 Proposal The original 20/80 proposal's dramatic switch in values did not last long. It succumbed to the concern expressed by the Office of General Counsel and the Office of Civil Rights in the US Department of Health and Human Services that age matching contravenes the requirements of the Age Discrimination Act of 1975. As subsequently amended, the 20/80 proposal retains 20% survival matching and allocates the remaining 80% of kidneys on the basis of modified criteria for calculating waiting time. The amended 20/80 proposal modestly promotes the value of efficiency and preserves the strong commitment to first come, first served temporal equality. The OPTN/UNOS Kidney Transplantation Committee adopted the amended 20/80 approach, and the new kidney allocation policy went into effect on December 4, 2014.[6]

Equal Opportunity Supplemented by Fair Innings (EOFI) Independently of the work of the Kidney Transplantation Committee, L. F. Ross and colleagues (2012) proposed an approach to allocating kidneys that uses consideration of age to enhance equality and efficiency concurrently without, they contend, discriminating on the basis of age. The two components of their approach operate sequentially. The first step is equal opportunity (EO), which is designed to give candidates of all ages an equal chance of receiving a kidney. The second step is fair innings (FI), which allocates the higher-quality kidneys to the younger candidates who are worse off because they developed end-stage renal disease (ESRD) at an early age and consequently have had fewer years of healthy life. Age groups are created for candidates, and age ranges are created for donors. To give each candidate an equal opportunity to receive a kidney, the number of kidneys allocated to each age group is proportional to the number of candidates in that age group. Then, starting with the kidneys in the youngest donor age range and moving through the consecutive donor age ranges to the kidneys in the oldest age range, kidneys are distributed across the candidate age groups, starting with the candidates in the youngest age group and moving through the consecutive age groups to the candidates in the oldest age group. Within the candidate age groups, kidneys are allocated to particular individuals primarily on the basis of waiting time on dialysis.

Unlike the two independent components of both the original and the amended 20/80 approaches, the EO and FI components are integrated. Embedding the FI component within the EO component produces two different kinds of equality—statistical (not temporal) equal opportunity and rectificatory opportunity. Moreover, integrating EO and FI produces a synergy of equality and efficiency because redressing the disadvantage of the younger candidates concurrently aligns the expected life spans of candidates and kidneys. The EOFI proposal strongly promotes both equality and efficiency.

The value of efficiency is understood in the same way in all these approaches—extending the length of survival of both candidate and kidney.[7] The value of equality, however, is understood in different ways and instantiated by different methods.[8] The diverse conceptions and methods of equality generate disparate approaches to making tragic choices about allocating kidneys.

9.1.4 Assessing the Approaches to Allocating Kidneys for Transplantation

Kidney transplantation reduces suffering and extends life. The value of efficiency serves the goal of limiting tragedy. The more years of life after transplantation that can be produced, the less the suffering and the fewer the years of life lost. The value of equality serves the goal of managing inescapable tragedy in the least offensive way, i.e., by deciding who suffers and who dies in the fairest possible way. The moral challenge for policy making is to design a mixture of methods that facilitates both values effectively, judiciously, and responsibly.

The previous allocation system in the United States promoted equality and neglected efficiency. It had another problem. The kind of equality promoted by the previous system—first come, first served—is common in our lives and easy to implement, but as practical and familiar as this temporal form of equality is, it does not fit the allocation of kidneys. Queuing is a fair, objective way of distributing some scarce goods, for example, lining up to buy tickets to a movie or concert. Queuing allows people to choose whether the effort required for, and the probability of success of obtaining, a scarce good are worth it. Queuing is attractive because it is a measure of the subjective value of a scarce good and because the ordering in a queue is ultimately the result of free, informed choice. But people do not choose whether and when they get end-stage renal disease (ESRD), and given the superiority of transplantation to dialysis, people do not choose to have a kidney transplant. Getting ESRD is a natural lottery with only losers. People with ESRD do not freely join a transplant queue; people with ESRD are put on a transplant queue by others.

In extreme situations such as famine, food can be dropped from an airplane or handed out from the back of a truck to those who have lined up. Tragic choices about who will eat and who will starve to death are dictated by urgency and desperation. Commonly food goes to the savvy, the swift, and the strong. The allocation of kidneys should not be predicated on misconceived assumptions about choice or necessity. People with ESRD are put on a queue because making them queue is an easy, simple, transparent, familiar method for allocating scarce goods. With ESRD, however, temporal neutrality is temporal arbitrariness. Putting people with ESRD on a queue does not respond to the relevant differences and magnitudes of their misfortunes, but doing just that is essential to managing this tragedy "in the least offensive way."

The original 20/80 proposal was a striking reversal of the previous system because it promoted efficiency and neglected equality. Although it was not itself a compromise between equality and efficiency, it would have been a major transition in the process of compromise as cycling through the two values over time. From that perspective the original 20/80 proposal, directly following the preceding first come, first served system, would have completed a cycle that gives roughly equal importance to equality and efficiency over the cycle.[9]

Calabresi and Bobbitt would understand the demise of the original 20/80 proposal because they emphasize that the emergence and fate of any approach to making tragic choices depends upon the nature and fundamental values of the society making the choice. The steadily increasing number of older people in the United States, a cherished tradition of constitutionally proclaimed and protected individual rights, a robust ethos of prohibiting and rectifying discrimination, and a ready penchant for litigation are formidable obstacles to age-based categorizations anywhere.[10] By retaining survival matching for 20% of kidneys and reverting to allocating 80% of kidneys on the basis of first come, first served, the amended 20/80 approach endorses efficiency only modestly, but sufficiently to be a substantive compromise between equality and efficiency and to contribute to the process of compromise as cycling. Nevertheless, its temporal equality, while workable and clear, is impervious to the nature of tragedy.

The EOFI proposal promotes two different forms of equality and concurrently and concordantly fosters efficiency strongly enough to represent compromise as cycling. The FI component especially befits the nature of the required tragic choice. Although the loss of any life can be tragic, the death of a child from leukemia seems more tragic than the death of a nonagenarian from cardiac failure. Similarly, succumbing to ESRD at a young age is more tragic than succumbing to it at a much older age. By giving priority to more tragic tragedies, FI limits the tragedy of kidney scarcity, as does EOFI's production of efficiency. In addition, the robust statistical equality of the EO component deals with the "irreducible minimum" of tragedy in "the least offensive way." Only a pure lottery, as might be used in drafting soldiers to fight in a war, would be a less offensive way of responding to ineliminable tragedy. And by simultaneously affirming relevant forms of equality and restoring efficiency, EOFI minimizes the need for compromise between those fundamental values. With its complementary, apposite forms of equality and its synergy of equality and efficiency, EOFI best fulfills the two criteria of limiting tragedy and dealing with tragedy in the least offensive manner.

But how does the EOFI proposal fare with respect to the charge of age discrimination that scuttled the original 20/80 proposal? Ross and colleagues hold that the FI component is "a nondiscriminatory use of age, provided that the number of kidneys allocated to each age group is held constant … because it treats individuals equally at different life stages," i.e., it is a form of prudential life span equity (Ross et al. 2012: 2118). They also point out that their EOFI proposal promotes efficiency not by allocating more than an equal share of kidneys to younger candidates but by allocating higher-quality kidneys to younger candidates, which uses age as a proxy for quality (2120). Those are explanations of why and how they are treating younger candidates differently. Because there are good reasons for treating younger candidates differently, EOFI does not discriminate "arbitrarily" against older candidates. But they are not a response to the arbitrary line-drawing objection that would be the impetus for litigation. Wherever and however EOFI draws the borders of age groups and age ranges, those borders will be vulnerable to legal action. In this regard allocating kidneys exposes the real fundamental value of North American society—not prolonging lives, not alleviating suffering, not recompensing the most vulnerable, but preserving and protecting individual rights.

Procedural values also are relevant to assessing approaches to allocating kidneys. An allocation system must maintain public trust and confidence, and to do that it has to be simple and transparent. It is no accident that methods of allocating kidneys are presented as "algorithms." The procedures for allocating kidneys have to be, and be perceived as, intelligible, objective, reliable, and immune to manipulation for the public to have faith in their legitimacy and fairness. The previous system and the original and amended 20/80 approaches are simple and easy to understand. The EOFI proposal is more complicated but not unduly so. The notion of fair innings is a comprehensible and appealing response, medically and morally, to the plight of younger candidates, and the nature of and reason for the equal-opportunity component can be readily explained. Given its many virtues, EOFI should not be dismissed because it is allegedly too complex or arcane.

Moreover, public trust and confidence depend on the process from which an allocation system emanates. The public dissemination of the revisions to the previous allocation system proposed by the Kidney Transplantation Committee (KTC) and reports of the meetings of the KTC on the Organ Procurement and Transplantation Network website, along with calls for public comments, are commendable. In addition, the KTC "implemented a detailed communications strategy which included a media webinar, tailored presentations for both professional and lay audiences, and documents to address frequently asked questions" (KTC 2012: 22). That "communications strategy" presents information about what the KTC is doing and how it is doing it. What is crucially missing, however, is an explanation of why the KTC is doing what it is doing. That omission was dismayingly evident in the KTC's insouciant rejection of the "theoretical" EOFI approach to allocating kidneys (KTC 2012: 22) Its summary of EOFI mentions only the EO component, ignoring the FI component entirely, and criticizes it for not providing equality for candidates disadvantaged by blood type or degree of sensitization, matters addressed by the KTC's own amended 20/80 approach.

The KTC's process of revising the system for allocating kidneys is designed to find a practical medical approach to making transplantation more efficient without violating carefully selected forms of equality. The KTC's orientation proceeds from its composition. Of its 28 members 20 are physicians from hospitals or medical centers. Four of the remaining eight members work in transplantation. Two others are from organ donation and transplantation organizations, and of the two General Public members, one is a doctor. No approach to allocating kidneys can satisfy all of the diverse forms of equality. How and why did these members of the KTC make their choices about which forms of the fundamental value of equality are most important to allocating kidneys and about how a compromise between those forms of equality and efficiency can be achieved in a revised allocation system? An approach to allocating kidneys of course has to be practical and grounded in extensive medical expertise and empirical research. But social choices about allocating kidneys are paradigmatic tragic choices, as discomfiting as that realization is, and have to be made by a process that is cognizant of, and responsive to, their essential nature. A process of compromise requires a more diverse and representative group of participants who are committed and empowered to defend substantive values and positions, sensitive to procedural values, and willing to make the concessions that are needed for an outcome that is a compromise at a time and over time.

9.1.5   Moral Compromise

One transplant physician said of the KTC's amended 20/80 proposal: "It's a compromise. I think it's going to make very little difference" (Sack 2012). What makes that proposal a compromise, moreover, one that would make little difference? The doctor who chaired the KTC described the conundrum for its members: "We want to maintain equal access and do better with this pool of kidneys. But by changing allocation slightly and getting 10,000 more life-years lived, what is that worth? Is it worth slightly decreased rates of access for certain groups of people? That is what we go back and forth trying to decide" (Sack 2012). Given that interminable vacillation, why not search for a compromise? Allocating kidneys has the hallmarks of a moral compromise. Neither equality nor efficiency can be satisfied fully, yet neither equality nor efficiency can be abandoned. How can each value be satisfied partially and proportionately to its importance in the circumstances?

In chapter 6 we examined in some detail Martin Benjamin's vivid example of how moral compromise could work in an intensive care unit. We saw that moral conflicts encompass not just discrete substantive issues, in that case whether to continue aggressive treatment for the patient, but the people who have to resolve the issues, the contexts within which they decide, and the processes they use to arrive at their decisions. Compromise involves a subtle but powerful and quite specific process of multiple compromisings that can lead to an outcome that is a compromise. Because the parties have to agree to a compromise, they all have the power to ensure that their position is represented reasonably and sufficiently in a compromise outcome. In addition, the participatory values displayed in a process of compromise enhance the legitimacy, rationality, and acceptability of the outcome.

Calabresi and Bobbitt acknowledge the place of compromise in designing an approach for making tragic choices when they characterize the melding of disparate methods "as a compromise between basic approaches usually entailing in modified versions some of the shortcomings (and concomitant powers) of each" (1978: 146). The mixing of different methods will be guided by the way the powers and shortcomings of the methods contribute to partially and proportionately satisfying the values that conflict in the circumstances. Compromise at a time is the familiar kind of compromise. Another kind of compromise partially and proportionately satisfies conflicting values by cycling through those values over time.

Calabresi and Bobbitt recognize two kinds of cycling. One form of cycling is shifting back and forth between second-order and first-order determinations to reduce tragedy to a socially tolerable level. This strategy exploits the interdependence of first-order determinations and second-order determinations: when the number of people who can get a scarce fundamental good becomes drastically low or the distribution of a scarce fundamental good becomes blatantly inequitable, pressure builds to increase the supply of the scarce good, and yielding to that pressure reduces the tragedy to an endurable magnitude. The other form of cycling operates within second-order determinations by shifting among methods of allocation that accord different priorities to the fundamental values of equality and efficiency.

From their historical survey of how the United States, England, and Italy have made tragic choices, Calabresi and Bobbitt conclude that the "most subtle" of the methods adopted by the three countries to avoid tragic results is the "cycle strategy"—the constant replacement of methodology, methods, and mixtures of methods. Cycle strategy, they explain, is inevitable because "it accepts the fact that society faces the paradox of being forced to choose among competing values in a general context in which none can, for long, be abandoned" (1978: 195–196). But the cycle strategy allows society to "limit the destructive impact of tragic choices by choosing to mix approaches over time" and thereby to reaffirm values that have been endangered (196). Not only does the cycle strategy preserve the values it engages, the forthright admission that a cycle strategy is being used affirms the value of honesty.

A single parent struggling to cope with the demands of caring for a child and holding a job is a more mundane, familiar example of how compromise can extend over time. A single mother might decide that because she has not played with her son much this week, she will leave work early, pick him up at day care, and take him to the park. Earlier in the week the demands of her job prevailed, but now it is time to give priority to her son. Subsequently, her job will reclaim priority, only to be curtailed again by the mother's love and devotion. That temporally extended compromise will continue persistently, but not always uniformly and sometimes unpredictably, for some time in the conjoined lives of mother and son. And over that time cycling will leave a residue of unsatisfied obligation, often expressed as regret and sadness in the mother. But because it is the best she can do, the negative emotions of remorse and guilt are inappropriate, since they focus the bearer on (unavoidable) failure and looking backward, as with Sophie and Jillian Searle. In contrast, the residue of the continuous compromising of the single mother seems focused on the unfolding, malleable future. A society similarly faces many tragic choices and can decide in response to their residues to look backward in recrimination or forward in hope, with determination to secure improvement.

In the concluding chapter of *Tragic Choices* (1978), Calabresi and Bobbitt explain that it is not the "business" of their book "to resolve tragic choices by means of discoveries of new methods, but to make it possible for us to get a clearer view of the state of affairs that troubles us" (195). Recognizing the two forms of compromise—compromise between equality and efficiency in a single approach to allocating kidneys and compromise between equality and efficiency via cycling through successive approaches to allocating kidneys—contributes to a clearer, more expansive view of the history and the current state of affairs with regard to this perplexing issue.

It also helps to make clearer how the entire historical period can be understood as an instance of core problem-solving process, similarly to the way we understand research in science (see chapter 4, section 4.3, ultimate paragraph, and references). The kidney era begins with scarce dialysis machines and ends with scarce transplant kidneys, accompanied throughout by a steady queue for the available service, with death and disability for those who miss out. The initial problem was how to allocate scarce machines to the queue, and the tacit solution criteria were something like a simple, familiar assignment rule that could be easily operationalized, explained, and defended. The initial proposed solution was the first come, first served rule or time-in-the-queue rule. This is an uncomplicated rule, easily administered through simple registration and familiar from many other queuing situations, from concert tickets to ice cream cones to toilets. The subsequent period saw an exploration of improved criteria for adequate solutions, expressed in increasingly sophisticated rules for kidney assignment, represented first by a shift from just one value to recognition of two primary values (equality and efficiency), and second by a two-tier conception of criteria, i.e., the introduction of first- and second-order determinations. This pattern characterizes most successive reformulations of solution criteria. But these immediately trigger correlative reformulations of the problem to be solved. Initial problem formulation would have been the (quite properly) relatively vague "find out how to assign services to the queue," but this ultimately becomes a much more focused "provide a rule for queue service that recognizes (a) conflict among multiple values and (b) the use of both first- and second-order determinations, and (c) is expressed in an administratively manageable procedure, and is relatively intuitively accessible to queue participants." An enormous amount of normative learning is manifest in these transitions.

Finally, along the way the methods for tackling this problem have altered accordingly. The initial time-in-queue rule required no more than simple addition to calculate its consequences. Those consequences revealed an intuitive flaw, the potential under the rule to waste scarce and costly resources on those with little expectation of using them, and indeed the prospect of emphasizing this flaw because those waiting longest will also have lived longest with the damaging disease. This is parallel to scientific research where often the initial methods stimulate critiques that had not beforehand been salient. So here too the search for methodological adequacy helped stimulate change throughout the whole process, which itself culminates not simply in the present sophisticated solution structure, but also in an appreciation of the whole landscape of scarcity problems of this kind—a recognition of what structural issues, like value conflicts and determination orders, are pertinent to developing adequate solutions to them and of how methodologies, problem formulations, solution criteria, and data are required for their solution. None of these changes would have been rationalized, and many simply would not have been comprehended, had the process not been backed by the resources and procedures of nonformal reason.

9.2   Designing a Process for Policy Making

Having examined in some detail the role of ethical design in the formation of a public policy that has developed and changed over decades, we turn to a recent initiative to intentionally design a rational deliberative process for creating a public policy in health care. Such an undertaking bolsters the importance our account gives to a process centered on ethical design, one whose rationality is based on nonformal reason.

Anthony Culyer (2006) provides an illuminating account of how the National Institute for Health and Clinical Excellence (NICE) in England and Wales designed a deliberative process for assessing the cost-effectiveness of health care technologies. A deliberative process, he explains, is "one that elicits and combines evidence of different kinds and from different sources in order to develop guidance" (317). Three kinds of evidence are relevant to the evaluation of health care technologies. One is context-free scientific evidence from research that controls for confounding variables to test hypotheses, for example, a placebo-controlled trial. Such research is conducted in clinics, laboratories, and centers of excellence and is supplemented by systematic reviews and meta-analyses. The second is context-sensitive scientific evidence from research that likewise tests hypotheses systematically but in a manner pertinent to the domain where the technology will be used, for example, a cost-effectiveness analysis of alternative drugs for treating a disease in general practice in England and Wales. The third is evidence that is neither scientific nor systematic, characterized as "colloquial." This kind of evidence is adduced on those frequent occasions when it is the only available evidence applicable to a particular issue, and it is used "to fit context-free scientific evidence into a context and to supply the best evidence short of scientific evidence to fill in any relevant gaps" (299). Colloquial evidence amounts to the professional judgments of experts. Given the complexity of the issues and the limitations and diversity of these three kinds of evidence, the guidance that emerges from a deliberative process is not the guidance of right answers but the guidance of informed, critically scrutinized rational judgment.

At the core of a deliberative process for assessing the cost-effectiveness of health care technologies is the challenge of amalgamating the results of technical analyses of clinical issues with the results of social-scientific analyses of the contexts within which decisions will be made, which requires combining not only the three disparate kinds of evidence but the plurality of values inherent to the technologies and implicated in their operational contexts. Culyer describes that endeavor as "a process of 'weighing up' and considering the contexts in which the guidance emanating from the deliberative process is to be used" (2006: 301). Unlike top-down decision making, this process involves stakeholders in that weighing up and examination of contexts, and laypersons are consulted and sometimes allowed to participate.

To understand this deliberative process, we need to understand the general nature of the issues, the goals of the process, and the design of the process. Features of the issues appear in twelve "conjectures," as Culyer dubs them, about the circumstances that make the use of deliberative processes "more likely" (2006: 301–302). What he means by "more likely" is, unfortunately, not clear. More likely to be tried or used? More likely to be apposite? propitious? helpful? successful? The list of conjectures is motley, but some of their points can be surmised. The first one, "decisions have been delegated by a body with a democratic mandate to one without it," could establish the legitimacy of the process and impose an ethos of democratic egalitarianism on the process. The last one, "wide public and professional 'ownership' is desired," would enhance the quality of the process and promote the acceptability of its outcomes. The other conjectures pertain to the substance of the deliberation. Several recognize the diversity and complexity of the empirical evidence: "evidence from more than one expert discipline is involved" and "evidence from more than one profession is involved." A number highlight the uncertainty and novelty in the evidence: "there are technical disputes to resolve and the evidence may be scientifically controversial"; "evidence gathered in one context is to be applied in another"; "there are issues of outcome, benefits, and costs that go beyond the conventional boundaries of medicine"; and "there are issues of implementability and operational feasibility involving knowledge beyond that of the decision makers." The remainder note that the evaluative dimensions of the issues are equally diverse, complex, and uncertain: "stakeholders have conflicting interests"; "there are other social and personal values not taken into account in the scientific evidence"; "there is substantial uncertainty about key values and risks that needs to be assessed and weighed"; and "there are issues of equity and fairness." The upshot is that judgment pervades the entire process, from the framing, to the analysis, to the guidance that resolves issues.

The goals of NICE's deliberative process are numerous and ambitious. Their procedures should be conducted with the greatest transparency and with extensive participation by "stakeholders," who are defined broadly as "patients, informal caregivers, clinical and other professional caregivers, health care managers, manufacturers, researchers, and the public in general" (Culyer 2006: 303). NICE seeks the respect of the clinical and research communities and the support of the Royal Colleges. It wants to avert dismissal of its work as "cranky, under-researched, or second rate," and it wants its guidance to be acceptable to the users of the NHS and fair to the inventors and manufacturers of technologies (303). It wants managers to regard its guidance as "do-able," and it wants skeptics and those who feel threatened to speak openly so that NICE can respond appropriately. Comprehensive aspirations for a comprehensive agenda.

How did NICE design the structure and the procedures of its deliberative process to accomplish its ambitious goals? In a way that Culyer takes to be "a model of deliberative process":

- there are open Board meetings that take place bimonthly around England and Wales (these are accompanied by public receptions and "Question and Answer" sessions)
- minutes are published on the NICE web pages before confirmation
- there is a Partners Council
- there is a Citizens Council
- the membership of the Technology Appraisals Committee is set broadly
- there are extensive consultation exercises throughout the appraisal process
- there is an appeals procedure
- there are consultative processes about process
- there are extensive liaisons with Royal Colleges, Independent Academic Centres, and National Collaborating Centres

- there is considerable joint working with NHS R&D and the National Coordinating Centre for Health Technology Assessment. (2006: 303–304)

NICE's deliberative process for assessing health care technologies is open, multidisciplinary, multiprofessional, and multi-institutional and incorporates, as Culyer puts it, "what is sometimes somewhat unfortunately referred to as lay participation" (2006: 304). It is widely consultative and collaborative in its quest for creative, critical help and attuned to its multitude of partners and audiences.

Culyer recognizes that judgment abounds in this formidable, complex deliberative process:

> The ultimate product of a deliberative process is guidance shaped by judgment—judgment about an effect of doing something, its size, the ways in which it is likely to be achieved, for whom, for how long, its cost in terms of the resources used that would otherwise have been employed in other ways to achieve other good things, and how worthwhile it is. (2006: 302)

Judgments have to be made about, for example, whether outcomes or costs might change when the results of research are transferred from the context in which the evidence was gathered, and whether the conclusions of trials in other countries or health care systems can be transferred. Moreover, such judgments will likely have to rely on colloquial evidence—"the judgment of practising physicians and managers engaged in the deliberative process" (Culyer 2006: 306). Evaluating the risks associated with uncertainty and the import of uncertainty are further matters of judgment.

In addition, as Culyer emphasizes, "There is no escaping the fundamental fact that the decisions and recommendations of a body such as NICE are laden with judgments of value" (2006: 308), starting with the basic notions of efficiency and equity, which, he notes, are axiomatic but ambiguous. The perspective from which an analysis is done proceeds from ethical judgments about what costs are acceptable and what consequences are relevant. A particularly troubling judgment for NICE is how value should be attached to the quality-adjusted life-year (QALY), the accepted measure of health-related quality of life. Should a QALY for an elderly person have the same value as a QALY for a child, for example, or should a QALY for a person with a debilitating chronic incapacity have the same value as a QALY for a person who is robustly healthy? Throughout the deliberative process the content of general concepts and values must continually be refined and contextualized. As Culyer trenchantly explains: "Simple-minded ethical talk about 'equality' in the health context is mere rhetoric unless one gets to be specific about equality of *what* and of *whom* and *for what*. There is a lot of unpacking that still needs to be done" (Culyer 2006: 316; emphasis in original). All that specification is a matter of judgment.

How is this judgment-pervaded deliberative process to be assessed? For Culyer, "The test ... of a deliberative process is whether the resultant judgment is (or will be) more comprehensively 'evidence informed,' better matched to the context of application, more efficiently implementable, and more widely acceptable to those affected by it" (2006: 302). Still more judgments, now about whether an outcome improves the deliberative process. That test accords with nonformal reason because improvement is the goal of nonformal reason. But where are the rationality and the morality of a deliberative process? Culyer suggests that a deliberative process is the kind of process Norman Daniels advocates for rationing scarce health care resources—a process that can "account for reasonableness."[11] Daniels realizes that there will always be reasonable disagreement about which principles are relevant to setting priorities in health care and what weight those principles should have. His alternative is to develop fair, publicly acceptable processes for making decisions about priorities. "The moral legitimacy of limits and priorities," Daniels holds, depends upon "how they are set" (2000: 1300). Accountability for reasonableness has three elements: transparency, appeals, and revisions. The reasons and rationales for decisions must be public, along with the constraints on the reasons to be considered relevant. There must be opportunities to challenge and revise decisions. And there must be a method for appeals. Accountability for reasonableness also educates stakeholders about substantive deliberation and fosters social learning about the limits of health care resources and the reasons that are relevant to making fair decisions. As Daniels explains, "we need a process that allows us to develop those reasons over time as we face real cases" (2000: 1301). A fair process legitimates the decisions that emanate from it and teaches us how decisions should be made. With its openness, its extensive collaboration and consultation, the participation of stakeholders and laypersons, the scrutiny of its processes, and its appeal procedure, NICE's deliberative process is indeed accountable for reasonableness.

Just as a fair process produces fair decisions, with nonformal reason rational processes produce rational judgments. The encompassing design of NICE's deliberative process—its manifold interlocking procedures and its rich plurality of participants and collaborators—is a haven for the resources of nonformal reason. Scientists and health care providers are trained to observe and hone that skill with experience. The other participants also are observers. Laypersons pay attention to signs of their own health and the health of those around them; manufacturers survey their employees and inspect their equipment; managers watch the behavior of their staff. All the participants will use formal and nonformal reasoning methods, regularly and unconsciously, but induction from past experiences and arguments by analogy across contexts are likely to be particularly apt. Engaging with people from different disciplines, professions, and institutions will spur new perspectives and ways of thinking and creative analyses and constructions. And comprehensive critical assessment will be especially acute and sweeping given the diverse backgrounds, training, specializations, and experience of the participants. NICE's deliberative process is reasonable and rational.

Culyer disavows any interest in the morality of the process, but the preceding examination of how kidneys are allocated for transplantation shows that his repudiation is both mistaken and naive. It is mistaken, fortunately, because the deliberative process he examines is riven with moral values, playing their inevitable part in the design of that process, as well as in its outcomes, which, as Culyer admits, "are laden with judgments of value" (2006: 308). Those values start with the basic notions of efficiency and equity, just as in the process of allocating kidneys. Moreover, ignoring the role of design ethics is naive because it neglects the importance of discovering the value structures of problems, what and how values conflict and what and how both first- and second-order determinations could address those conflicts. Again fortunately, though, Culyer acknowledges that the perspective from which an analysis is done proceeds from ethical judgments about what costs are acceptable and what consequences are relevant, hinting at an awareness of the importance of exploring the values structures of providing technologies.

Culyer observes that few organizations have undertaken such a daunting challenge "quite so boldly and imaginatively" as NICE has (2006: 299). In the title of his article, he commends NICE's approach for being "an exemplar of a deliberative process." The success of NICE's work of course remains to be seen. Nevertheless, its deliberative process is impressively designed, and the rationality of its design will be imparted to the policies it designs.

9.3 Conclusion

The examples of policy making in this chapter illustrate and vindicate two crucial features of re-reasoning ethics. One is the ubiquity, the necessity, and the flexibility of compromise. The principle-based ethics of formal reason has no place for compromise. Nonformal reason recognizes that values suffuse ethics and conflict, and nonformal reason provides the rationality of the deliberative processes that critically and creatively resolve conflicts between values, for instance, the conflict between equality and efficiency in devising methods of allocating kidneys for transplantation. The other feature is the inevitability of judgment. The NICE model of a deliberative process for assessing the cost-effectiveness of health care technologies recognizes that conceptual, empirical, and evaluative judgment is inescapable and pervasive, and so it designs the structure and the operation of the process to be broadly participatory and extensively collaborative and consultative. The design of NICE's deliberative process proceeds from the appreciation that judgment is inescapable and the realization that the social implementation of the resources of nonformal reason can make judgment rational.

With chapters 8 and 9, dealing with the designs of institutional processes (already raised in chapter 7) and public policy, we have begun to open up the scope of ethical design, moving from what Avishai Margalit calls micro-morality—the study of interpersonal ethics—toward increasing macro-morality, the ethical issues in the designs of professional roles, economic systems, political institutions, and culture proper. These are all venerable and worthy subjects, calling for their huge contributions to the design of human flourishing to be elucidated. This opening of the scope of design forms a natural, unified extension of the notion of design for flourishing.

Notes

Acknowledgments

1. © (2013) The Milbank Memorial Fund

Chapter 1

1. For accounts of the life of Joseph Merrick, who became known as the "Elephant Man," see Montagu 1971 and Howell and Ford 1980.

2. See Lippman-Hand and Fraser 1979a, 1979b, 1979c, 1979d, and Lippman-Hand 1982.

3. This account of the work of Lippman-Hand and Fraser is an elaboration of the discussion in Hoffmaster 1991, as refined and applied in Hoffmaster and Hooker 2009.

4. How these women frame their problem is not idiosyncratic. In his account of living with cancer, Diamond describes his reaction to the 1 in 100 chance that a fetal blood test on the baby his wife is carrying could induce a miscarriage and the 1 in 250 chance that it would indicate a chromosomal abnormality that could cause the baby to die of asphyxiation as it is being born, and then turns to his own chances of being cured: As odds went they were reasonable ones and had I been standing in the bookie's thinking about putting money on a healthy baby there would have been no problem. But under the circumstances the odds were meaningless. In these situations all odds are the same odds, or rather evens. Either the worst will happen or it won't: the baby will die or it won't. The odds always feel like 50:50.

And as it was then so it was for me now. Either they'd cure me or they wouldn't. Everything was 50:50. When the official odds were 92 per cent they were really 50:50, and if the operation didn't work and the official odds dropped to 20 per cent or 10 per cent or worse, they'd still be 50:50. (Diamond 1998: 138–139)

The meaning and significance that people attribute to the "official odds" are highly contextual; this is an expression of risk averseness in the face of a serious risk, a common position. But that kind of value assignment is impossible under the standard MEU model, for which the value of an outcome remains constant irrespective of its probability of realization. This is called the risk-neutral position and is another consequence of the fact/value dichotomy.

5. Women who remain torn, strongly desiring a child but unsure of their ability to handle the worst they can presage or to accept responsibility for a decision they might regret, not infrequently resolve their conflict by engaging in "reproductive roulette," that is, they use contraceptive measures they know are insufficient. They relegate the decision to "fate." How to respond to this conflict is a further, separate issue beyond that of employing scenarios to achieve the decisive go/stop formulation that serves most women. In this chapter we examine only the lessons of the antecedent decision.

6. We make no claim that the women's decision-making processes are either intelligent or rational in every detail. Humans do not always use the most effective problem-solving strategies, and there are many biases, illusions, fallacies, and the like that infect the methods we do use. The women's strategies undoubtedly are affected in some of these ways. These complications eventually have to be taken into account in assessing precisely how rational the women's actual problem-solving processes are and, more importantly, in helping to improve them, but that task is not germane to our concerns, namely establishing that a very different (from rule-following) *overall* approach to problem solving is eminently intelligent, so rational. Imperfections that can occur in the processes of that approach do not, and cannot, disturb our conclusion. Moreover, the discovery and rectification of such imperfections in science are themselves triumphs of the kinds of rational problem-solving processes identified here. For further discussion, see chapter 3.

Chapter 2

1. As naturalists, we do not find the unclear borders of moral norms a problem per se. Rather, that boundary becomes a fallible theoretical posit within a fallible, learnable normativity that can be improved as human experience accumulates. See chapters 3 and 4.

2. For Beauchamp and Childress, though, the four principles provide only a "framework" for addressing moral problems and "do not constitute a general moral theory" (2001: 15). The eventual development of a moral theory is prompted by the need to specify and balance principles.

3. We are grateful to Dr. Dori Seccareccia for providing this vignette and for giving us permission to use it.

4. One definition of palliative care is that it is "the standard of care when terminally ill patients find that the burdens of continued life-prolonging treatment outweigh the benefits" (Quill, Lo, and Brock 1997: 2099).

5. This vignette is extracted from Marshall 2001.

6. For a more dramatic example of a mismatch between a patient's approach to moral reasoning and a clinical ethicist's approach, see Churchill 1999: 256.

7. For a clear and compelling account of how a clinical ethicist, not working within the confines of the applied-ethics model, can be helpful, see Zaner 1993.

8. This vignette is extracted from Paulette 1993.

9. Llewellyn (1960) illustrates this precise problem in judicial decision making with his list of "thrusts" and "parries" that conflict in the interpretation of the law. See his appendix C, "Canons of Construction."

10. *Saskatchewan (Minister of Social Services) v. P. (F.)* (1990), 69 D.L.R. (4th) 134. References to the opinion of Provincial Court Judge Arnot are given in the text.

11. This vignette is taken from Arras 1988.

Chapter 3

1. Note that, secondarily, ethics also applies epistemic values to ethical methods because epistemically valid ethical understanding is sought, just as science applies ethical values to epistemic methods because ethically valid epistemic understanding is sought. However, these additional relationships do not change anything of importance here.

2. For the party example, see Hatchuel 2001. The criteria for determining what defines a "nice party" are not well-specified and have to be discovered as part of the problem solving process. The party location could include a home, night club or rental hall, or virtual locations. Location will partly constrain the options for entertainment, which may include party games, dancing, conversation, theater, or sport. Neither are the criteria of success well defined: is it a nice party if the guests stay long or instead drift in and out relaxedly? Should they express appreciation for the memories, or be unable to remember anything?

3. A working through of the many theories involved in one well-known scientific instrument, the cloud chamber, is presented in Hooker 1975, reprinted in Hooker 1987a (chapter 4, appendix 2; n.b.: interchange C_4 and C_5 labels in the diagram).

4. See Brown 1988. Formalizing remains an active pole, leading to splitting off of fragments of these nonformal skills and trying to characterize aspects of them more explicitly, obtaining metaphorical/analogical and causal reasoning and the like that can satisfactorily be made more formalized for some sufficiently circumscribed but still useful contexts (e.g., in engineering design problem solving). But the contextual decision that selected aspects of problem solving in the subdomain have been adequately captured itself requires many nonformal supporting judgments, and anyway

contextual and historical-developmental individuality of skilled judgments always outruns any such fragmentary achievements.

5. BH was doing ethics rounds and consultations at several hospitals and health care institutions in the 1970s and 1980s, and he was a member of his university's REB (Research Ethics Board) for eight years. CH was chair of his university's research ethics committee at a time, the late 1980s, when such IECs (Institutional Ethics Committees) were still largely being invented. As members of those committees, we learned about their roles, and we learned in our roles as members of the committees. Our experiences taught us the importance of the issues at the core of this book. In particular, we realized that if ethics were a matter of applying a priori moral principles and rules, the job of the ethicists would merely be to educate the doctors, the researchers, and the community representatives on the committee as to the principles and valid inference—like the lawyers on the committees who explain health care law.

6. For some of the history of method, see, e.g., Blake et al. (1960), Brown (1988), Oldroyd (1986), and Shapere (1984); the latter explicitly connects it to the development of scientific rationality. This developing characteristic of method is well displayed in the development of logic itself and especially its explosive exfoliation this century (modal, probabilistic, many-valued, fuzzy, … logics). For contemporary disputes about methodological procedures, see, e.g., the disputes about statistical inference (e.g., Harper and Hooker 1976 and Breiman 2001; cf. G. Hooker and C. Hooker 2017, Hooker 2011c, section 6.2.5), and for the methodological consequences of the rise of complex systems models across science, see Hooker 2011a.

7. For more on the shaping/shaped character and dynamics of culture, see Hooker 2002.

8. Briefly, confirmation is the measure of the degree of positive support that a body of evidence gives to its explaining hypothesis (or hypotheses). Empiricism is the normative methodological requirement that rational science ought to accept only those claims that have sufficient evidential support and seek to maximally increase this degree of support. Falsificationism is the normative methodological requirement that rational science ought to accept only claims that are falsifiable and ought to test hypotheses in ways that maximally increase the likelihood that they will be falsified if they are untrue in any respect. For references see Brown 1979, Hooker 2010, (references, note 13), or any standard textbook. These norms typically impose conflicting requirements on method. Again, foundationalism is the normative methodological requirement that rational ethics ought to deduce its right actions from the right fundamental normative principle that applies, while coherentism is the normative methodological requirement that rational ethics ought to derive its right actions from a sufficiently (but typically not wholly) coherent set of judgments about applicable principles and values. These norms also typically impose conflicting requirements on method.

9. An argument "P_1 and P_2 and ... P_n, so C" has as its equivalent form "If P_1 and P_2 and ... P_n, then C." The argument is valid, that is, necessarily preserves truth from premises to conclusion, if and only if its equivalent form is a logical tautology, that is, is true by its form alone. This provides a sharp normative formulation of the bounds of rationality. However, its elegance is spoiled in practice, since any logic that is complex enough to model at least certain fragments of arithmetic is undecidable and thus it may be undecidable whether an argument form is a tautology of that logic. Deciding the matter involves constructing a proof, and doing that is then a matter of skilled judgment (and perhaps luck)—and is thus a rational, certainly an intelligent, activity that itself escapes this narrow criterion of rationality.

10. In the case of science, if pure observation is considered also error-free and scientific method confined to confirmatory logical inferences from pure observations, an elegant (if unconvincing) case is completed for the error-freeness of science. Even if, as with Popperian falsifiability, the body of hypotheses "in play" at any time may not be error-free, the refutation of those containing error, on the basis of empirical data, is error-free. Similarly, in ethics if factual characterizations of situations are presumed error-free then right action deduced from certainly right normative principles is again error-free.

11. If the premises of two deductively valid arguments are conjoined, and their conclusions likewise, then the argument from conjoined premises to conjoined conclusions must again be deductively valid. In this way the scope of deductively valid conclusions can be consistently extended. This is not true of inductive arguments, essentially because there can be data sets D1 and D2 supporting conclusions C1 and C2 respectively such that though D1 and D2 may be consistently conjoined, C1 and C2 cannot, contradicting each other. The same outcome can be had with a deductive and an inductive argument—see Hooker 2010 for details and note 13 below for one class of examples. The problem is widely recognized, e.g., leading Levi (1980) to acknowledge induction's "myopia" (localness) and Carnap to qualify inductive scope with the requirement that all the available evidence be considered (total evidence requirement, Carnap 1947), but whether Carnap's requirement is satisfied cannot be formally decided. Even if there may be some limited valid arguments of inductive form (see Stove 1973), this cannot relieve the present limitation.

12. See, e.g., Hooker 1995 on Bertrand Russell's rational (but unlucky) chicken, where D describes sequential daily care from January 1 to December 23 and C asserts that care will continue indefinitely, I specifies that it is December 24 and R that on December 24 fattened hens are killed for Christmas dinner. Then we have "If D then C" but "If D and I and R then not C," a change of argument form brought about by consistently conjoining just two data to D. It will be seen immediately that such situations occur, or threaten to occur, throughout science whenever we have partial knowledge of processes.

13. This is problem enough if T is internally complex but is made worse if, as is typical, many different theories are drawn on to make the prediction so that, in effect, $T = T_1$ and T_2 and ... T_n, for now errors may be located in any one or more of these. In many actual refutation situations in science the nature of the theory-evidence relation (e.g., whether causal, probabilistic, or coincidental) takes the form of a confirmed hypothesis, which only compounds the logical problems. The general ideas here are well known; see, e.g., Duhem 1962, Campbell 1957, Brown 1979, 1988, Galison 1987, and Hooker 1995 chapter 3, Hooker 2010, among many. The plasma probe example of section 3.1 above instances all these complications and is worked out in moderate detail in Hooker 1975 (reproduced in Hooker 1987a, chapter 4). When it comes to identifying errors, as Bickhard (1993, 2005) correctly points out, to know that an error has occurred requires only a judgment that a failure of anticipated (e.g., predicted) outcome has occurred. But to correctly *locate* the source and nature of the error requires a much more sophisticated understanding of the action situation and discrimination of the factors required for successful action in it. The young cheetah, e.g., will know, simply by her continuing hunger, that her hunt has failed. But learning to discriminate wind direction, camouflage, terrain kinds, and so on as factors in successful hunting and hence also as sources of hunting errors requires much more than this; it requires all her learning through to mature independence. See further chapter 4 and Christensen and Hooker 2000a, 2000b, 2002, and for a detailed scientific application (ape language research), see Farrell and Hooker 2007a, 2007b, and especially 2009.

14. See, e.g., Lakatos 1976, Shapere 1984; cf. Brown 1988 (p. 87): "all the early formulations of the substitution rule for the predicate calculus have been reported to be unsound."

15. Consider the three-way interaction now developing apace among mathematics, physical theory, and normative formal structures (logic, statistical inference, and formal problem solving), e.g., the development of quantum logic on the normative side and generalized theories of dynamics on the scientific side; of new foundations for probability and statistical inference on the normative side and new sampling and other instruments on the scientific side; of new mathematical theories of problem-solving / artificial intelligence procedures on the normative side and psychological and cybernetic theory and practice on the scientific side.

16. A formal MEU model of rationality also will not provide the requisite richness to rationality.

17. Involuntariness per se is necessary but not sufficient. Not every aspect of what is pushed involuntarily on us conveys truth. Many perceptual illusions and hallucinations, e.g., cannot be voluntarily corrected. It is indeed an example par excellence of the exercise of reason to learn to distinguish what respects of what involuntary signals reliably convey information. This issue is finessed by the foundationalism of empiricism (and rationalism—see note 19), but quite unconvincingly. It is avoided

by Popper's fallibilism only at the expense of indefinitely enlarging the dimensions in which potential errors are to be investigated.

18. Extreme instrumentalism is summed up in the Humean dictum: reason is and ought to be the slave of the passions. This reduces norms to conventions. Both empiricists and Popperians source their empirical knowledge in contact with this world but offer an ambiguous account of normative knowledge, being explicit conventionalists and tacit rationalists. Popper is instrumentalist/conventionalist in that the adoption of the maximally informative ideal for science is a nonrational decision (and an ethical one), but he smacks of rationalism in the way that he then tacitly constrains rational process to deductive logic as the foundation for falsificationist method. Conversely, empiricists may either deny having explicit ideals or regard them as conventional decisions, and thus support instrumentalism, but similarly smack of rationalism by accepting a priori that logic expresses rationality and thus taking truth, which logically valid argument necessarily preserves, to be the sole rational goal of science.

19. Accepting the skeptical assumption also requires a content foundationalism and, for formalism (logic), also a rule foundationalism (cf. Brown 1988). Philosophers like Popper are fallibilists about content and reject this aspect of the skeptical assumption. Yet they seem to accept foundationalism for reason itself. This is presumably because logic itself seems so clearly to require, and satisfy, rule foundationalism. Why settle for less? Again, this shows the hold which the formalist account of reason has.

20. This is well illustrated by Quine's flawed account of naturalism, which turns out to tacitly assume an empiricist metaphilosophy. See further Hooker 2010 (following 1995, chapter 6). Quine's paper does draw attention to the idea that in a naturalized conception the criteria of cognitively justified action merge much more conspicuously into the general criterion of rational action.

21. Lakatos, who expanded Popper's falsificationist conception of scientific method, considered that if a core scientific research program had sufficiently often to be defended from its otherwise-falsifying failures by adjustment of external assumptions, it should be evaluated as degenerating and be abandoned, to be replaced by another core conjecture. In that spirit, and taking rationality theory as ultimately in the same position as scientific theories, the core formalist program for capturing rationality should also now be regarded as a degenerating research program, and another substituted in its place.

22. The 24 values are made up of the width, precision, and accuracy of each of empirical adequacy, explanatory power, unity, and technological applicability, yielding 3 × 4 = 12 values (width of empirical adequacy across situations; precision of empirical adequacy, i.e., how finely quantified the predictions; accuracy of empirical adequacy, i.e., how small the error bars around agreement with experiment; etc.), each pursued either boldly (Popperian falsification) or conservatively (Carnapian induction), yield-

ing $12 \times 2 = 24$. Note that unity involves, but is not reducible to, explanatory width (whether applied inter- or intra- theoretically), and that other potential, but unlisted values, e.g., testability, are taken to be determined by these 12.

23. Material in section 3.3 is an improved, adapted version of the corresponding sections from Hooker 2010, with that on naturalism in turn improved and reworked from Hooker 1987a, section 8.2.2 (see also 1995, chapter 1) and that on finitude in turn improved and reworked from Hooker 1994a.

24. This ontological characterization harks back to the British naturalists, especially Alexander and Case, but also takes nourishment from the more epistemological tradition of the later American naturalists, especially Sellars (père).

25. Naturalism is concerned with unity of understanding, but it is not inherently concerned with promulgating any narrow materialism. In a nonlinear dynamical world emergence is pervasive. Conversely, as we have penetrated more deeply into the world's constituents they have grown more complex and strange, not less, until we now face the unlimited complexity and nonlocality of relativistic quantum fields. The nature of the world is their nature and, whatever that is, the old materialisms and immaterialisms alike look simplistic. The naturalist commitment is to understanding ourselves as one with nature, as differentiated from within a common natural framework, and to seeing our capacities as differentiated capacities within that framework.

26. Among the many further consequences is that any concept introduced should "grade back" sufficiently smoothly across the evolution of complex life forms to simpler physicochemical conditions. Physically, for instance, hormone regulation grades back from its manifold mammalian regulatory roles to elementary rate regulation of various aqueous reactions.

27. This necessity is no defect; rather, it is a consequence of our adaptability—our chief strength, given the limits of broad biomolecular finitude, because it is more powerful and resource efficient than more explicit storage of options.

28. This account is given a preliminary development in Hooker 1991, which has its roots in turn in Hooker 1987a (section 8.3.2 and elsewhere), and given fuller exposition in Hooker 1995 (chapters 5 and 6).

29. For instance, if exploring value-increasing bank branch designs leads to methods for spreading banking functions across several floors instead of just the ground floor while providing more convenient upper-floor contact with related services such as insurance and project management, and ground-floor community services such as children's play area and parking, then a whole new realm of design possibilities has been opened up in methods, problem and solution conceptions, and so on, highly indicative of a successful problem-solving approach. For an explanation, see Hooker 2017a, 2017b.

30. See Hooker 2010 (following 1995, chapter 6), in response to Putnam 1982.

31. Putnam's argument parallels Moore's argument for a nonnatural account of goodness, and Moore's argument fails for similar reasons. And similarly for any norm. Indeed, the naturalist response here applies to any irreducible theoretical conception, not just to reason. With the exception of C_3, these are characteristics required for any irreducible feature of life about which we are able to learn; they are not special to reason. And since reason is a basic cognitive capacity about which we are able to learn, our conception of it continues to grow and refine, so that no definition of it in current terms will be completely satisfactory. But this is again a feature shared by all concepts, certainly by all theoretical concepts. Under the attractively naturalistic Peircean conception of concepts as open-ended dynamic constructs being constantly adjusted to suit their evolving roles in evolving theories and practices (Legg 2005, cf. Brown 2007), every explicit characterization of a theoretical concept is tentative.

32. For further detail, see Cherniak 1986, and for its further development, see Hooker 1994a.

33. For elements of this story, see Bickhard 2002, 2005, Hooker 1987b, 1995 (chapters 5 and 6), 2009, 2010, 2017b. In this book we show how this approach to rationality works out and the multiple and deep differences it makes to ethics in turn. But extending the account to indicate how they fit into a naturalist theory of human agency generally, how they can be embedded in a broader biological setting as natural phenomena, and how they have been concretely, historically implemented is largely beyond the scope of this book.

34. Hooker (1994a) has labeled these *degenerate idealizations*: idealizations, meaning that some content has been omitted in order to simplify the account; degenerate, meaning that the collapsing out of content leads to irreversible change that cannot be reversed by simply re-adding what was omitted. Other idealizations are reversible simplifications. For instance, simplifying gas molecules to points is a reversible simplification, and shifting from relativistic to Newtonian space-time is a degenerate idealization. A large part of the history of fundamental physics is an uncovering of degenerate idealization where none was thought to be. We hold that the same is true for the rational foundations of ethics.

35. The classic account is Tversky and Kahneman 1974; cf. Kahneman, Slovic, and Tversky 1982.

36. See, e.g., Arpaly 2003 and, from a different angle, Misak 2008. While Misak argues for the enrichment of narrative in deliberation, Arpaly considers that these complexities require a new approach to ethical decision making, one that eschews deliberation and judgment. However, we find that he ultimately relies on some notion of reasoning, since he speaks of having reasons to act, but without providing

any account of this reasoning. Our book, by contrast, can be read as a detailed exposition of just how reasoning is done.

37. See Hooker 2010, section 5.

38. This formulation builds in two dimensions to the ideal: the completeness of the component tools available (e.g., the completed development of statistical inference tools) and their unrestrained use. The "all" here is intended to include not only the observational data but also the presuppositions of the observation regime, and so on, plus the decisions to terminate each of these inquiries and to commence new inquiries, and so on.

Chapter 4

1. The category of importance is that of skilled informal judgment. Formal judgments issue from formal deliberation and are thus necessitated; they are judgments in name only. The logic-based analytic ideal is to obviate informal judgment.

2. See, e.g., Professional Judgment Resource, Center for Audit Quality (www.thecaq. org) and Wedemeyer 2010.

3. A prerequisite is that there be a competent internal process of deliberative judgment formation, however elementary. A female mosquito compares internal indicators of sufficient (for reproduction) and actual capacity for healthy egg laying in judging whether or not to search for a blood host, but once a search is initiated she is a CO_2-gradient-tracking automaton. See Christensen and Hooker 2000a, Moreno and Mossio 2016.

4. Here Harold Brown's orientation returns us to agent capacities that underlie Aristotle's ethics as a useful starting point for understanding rationality and its practical application in ethics and law. Thanks to Brown for the following summary. Three concepts from Aristotle's ethics—deliberation, practical wisdom, and equity—will help clarify the notion of judgment. According to Aristotle, deliberation is the ability to arrive at reasonable results in situations in which we do not have the grasp of necessity that he holds to be characteristic of science, but in which we are also not totally ignorant. Deliberation is particularly important when we are concerned with human affairs, where the circumstances and the possibilities are too complex to be captured in a set of usable explicit rules. The ability to deliberate well is the central characteristic of what Aristotle calls "practical wisdom." This is the ability to arrive at fallible but nonarbitrary decisions "about what sorts of thing conduce to the good life in general" (*Nichomachean Ethics*, 1140a, in McKeon 1941: 1026). Ex hypothesi, this ability is not exercised according to rules, but by specific individuals who have developed practical wisdom as a result of their experience of human life. The third concept, equity, is the ability to override an established rule in order to deal with the special features of a particular case (a striking example of the need for deliberation of

a particular kind). Aristotle was concerned with exceptions to legal rules in situations where the law, exactly because it must be expressed in universal language, gives a clearly incorrect outcome in a specific case. When situations of this sort arise, to deliver equity we turn to those who can exercise practical wisdom. We depend on them "to say what the legislator himself would have said had he been present, and would have put into the law had he known" (*N.E.*, 1137b, in McKeon 1941: 1020). However, Aristotle's insights can be generalized; we can develop the ability to deliberate and exercise judgment in many different fields, and it is this ability that we draw on when we run into the limitations of our understanding as we have codified it to date, including current rules, and wish to improve upon it.

5. In these cases pursuit of rule models would quickly lead to an implausible unending regress of higher-order or meta rules to cover required context-dependent changes in rules (cf. Aristotle on legal equity, note 4). In many cases there are in fact some rules for generating a skillful performance—for example, rules for good chess play or style guides for writers—but these rules are notoriously insufficient to generate a skillful performance. In fact, it is just at the point where available rules cease to be sufficient that differences of skill become particularly apparent. While every step of a valid mathematical proof is rule governed, constructing proofs is not; rather, there is in general provably no sufficient set of rules for deciding what step should be made at any given juncture. As chess and logic illustrate, even activities that are defined by rules can be, and often must be, carried out apart from any use of rules. (Current computers do not use rules, or make judgments, at all; they merely conform to rules.)

6. Note that support for a supra-rule rational capacity does not follow from support for a supra-rule performance capacity. Among the best-known instances, Dreyfus and Dreyfus (1986, Dreyfus 1991) have developed an account of physical and cognitive skills that has a rule-transcending orientation in common with the conception developed here, but they actually support a narrow, rule-following conception of rationality, despite their insistence on transcending it. They introduce a progression of expertise from novice and advanced novice to competent, proficient, and expert, the earlier stages clumsily rule-bound, the later stages increasingly fluid and powerful because skillfully intuitive and not rule-bound. Describing behavior that is contrary to reason as "irrational," they note that a "vast area exists between irrational and rational that may be called arational," and they conclude thus: "Competent performance is rational; proficiency is transitional; experts act arationally" (Dreyfus and Dreyfus 1986: 36; cf. Dreyfus 1991: 186). They speak, for example, of an expert, faced with a sufficiently new situation, being forced to turn to "detached reflection" and to "appeal to principles" (i.e., rules) (Dreyfus 1991: 247), rather than utilizing the creative construction of new criteria and actions his expertise provides (cf. radically new scientific research domains, like quantum theory in the early twentieth century). Contrarily, according to our approach, expertise, although not rule-bound, is a fundamental part of rationality.

7. As noted in chapter 3, note 14, Brown (1988: 87) remarks that "all the early for-mulations of the substitution rule for the predicate calculus have been reported to be unsound." See also Lakatos 1976. The logician Quine demanded consistency as a minimal rational requirement, all the while unaware that his own proposed set theory was inconsistent. Even consistency proofs for formal systems are not better founded: they are all relative to arithmetic consistency, but about it we have no consistency proof. See also Hooker 1998.

8. Thus we talk both of "grasping the point" of a spear and of an argument, of "gaining an overview" of both the countryside and the calculus, and so on. Much of even our abstract mathematics, e.g., vectors, gradients, and integrals, clearly extends spatiotemporal metaphors while other components extend action metaphors, e.g., operators and mappings, and it is at least plausible that all of our abstracting is of a similar kind.

9. It follows from this and the preceding discussion that "interpretation" here is a much richer notion than just that of an inner interpretation process corresponding to (c). See, e.g., Heelan 1988, Heelan and Schulkin 1998.

10. See Galison 1987 for a detailed study of one such scientific situation. As this discussion shows, the proper integration of observation into science and ethically salient practice depends on manifold judgments. Judgment enters into decisions about whether to pursue an issue (is the epistemic value it might deliver worth the risks and resource investments required?), what further observations to carry out, what means to use for carrying out these observations and under what conditions and with what instrument settings, etc., and finally, when to suspend further critical assessment and announce an observation as established. Nonetheless, in judiciously judged ways established observations are used to improve methods, theories, and practices, on which subsequent judgments are based, including those theories used in understanding our instruments of observation, those practices used in operating the instruments, and those methods used in processing their information.

11. For more on the theme of science as a dynamic system of judgments and prac-tices, and the same for reason itself, see Hooker 1995 and Brown 1988, 1996, 2006.

12. See, e.g., Carter and Carter 2005, Wikipedia.org, "Ignaz Semmelweis."

13. See Mill 1843/2002, Mill and Nagel 1950.

14. Each of these forms of reasoning has only slowly developed its contemporary form. With argument validity, Aristotle's syllogistic logic was the paradigm until the nineteenth century, although it is a very small fragment of modern logic; and logic continues to develop, partly as wider applications are sought (e.g., to counterfactual conditionals), partly in interaction with mathematics through its structural connec-tions with geometry, algebra, and computation, and partly in interaction with sci-ence, e.g., with the development of quantum logic (see note 17). In short, its

development looks like that of any other fallibly learnable theory. The progressive discovery of the logical paradoxes (Gödel, Lowenheim-Skolem, etc.), combined with the fact that proofs of soundness and completeness are not absolute but always relative to some intuitively correct infinite theory (typically, arithmetic) that already outruns our intuition, illustrates the small reach of our conceptual grasp on deep structure and should arguably leave us wary of making claims to infallible knowledge. The story is similar for the other forms of reasoning, only still more recent.

15. Logic is least effective in dealing with global properties; reasoning from a large set of interconnected premises quickly overwhelms our capacity, and notions like global system consistency quickly escape our formal control, as Cherniak 1986 (and Gödel's theorem) shows (see chapter 3).

16. Notice how, once a fallible process conception of rationality is adopted, logic can be evaluated from a wider perspective than just the necessity or otherwise of its inferential relations, with rather different outcomes for its role in, and centrality to, intelligence.

17. See Hooker 1973, 1975–1979 for early papers, and the many texts since. Putnam (1969) has argued that the relation between quantum theory and logic parallels that between general relativity and Euclidean geometry: In each case a formal structure that was long considered definitive (respectively, classical logic, Euclidean geometry) is challenged as a result of the continuing attempt to develop an adequate overall account of the world (respectively, quantum mechanics, relativity theory).

18. Finite, fallible agents typically cannot in practice maximize: ignorance of the future (including of how methods and utilities may develop) and resource constraints on both modeling the world and checking for errors, especially for complex situations, make it impracticable. The next best aim is any outcome of sufficiently satisfactory utility (see, e.g., Simon 1947/1997), but in the face of deep uncertainty even that must be reconfigured to any outcome of sufficiently resilient sufficient utility. See Brinsmead and Hooker 2005, 2011.

19. For this general conception see Hooker 1995, its roots in the discussion of goals in science, Hooker 1976, of belief as commitment, Hooker 1987b (especially section 8.3.2), and of rules versus utilities in science see Hooker 1981, 1982a, 1982b. For the systematic development of the idea of the scientist as risk-taking investor, including the emergence of epistemic rules, see Shi 2001 (esp. chapter 8, note 17 and text).

20. This is illustrated in the intelligent response to Prisoner's Dilemma games; see chapter 8. Modeling these "meta-game" choices as further formal games faces both regression and issues such as interactions among players with incommensurate game models (e.g., the exquisite treatment of Othello and Desdemona in Rapoport 1960; see briefly chapter 8).

21. See, e.g., Schon 1967.

22. Again, socially organized construction can be used to distinctively, and typically more powerfully, improve individual construction judgments. This is because socially organized construction (a) can implement construction processes that are unavailable to most individuals, e.g., complex models of scientific or ethical situations, (b) can be more discriminating than individual construction processes through widened skill range or by focusing more construction resources on a problem than any individual could provide, and (c) is more able to correct errors of construction than can individual checking processes by bringing to bear the foregoing methods. Well-designed individual and socially organized construction processes strongly interact in ways that can synergistically reinforce improvements in both. (And if poorly or maliciously designed, they can equally systematically undermine both.).

23. Calabresi and Bobbitt 1978.

24. It used to be obvious that ethical practice was not as well organized socially as was science. Today, however, we face the increased sucking of science into the marketplace as private businesses conduct useful scientific research, while the roles of ethicists in those same businesses, as well as in public institutions (including those of science), continues to grow and become visible. The outcomes remain to be seen.

25. For instance, there is a geneticist's understanding of a plant's variational capacities so thorough as to identify a new kind of genetic shift far ahead of theoretical ability to understand it, beautifully expressed in Keller's essay "A Feeling for the Organism" (1983); the ability of Scott-Payne, a founding contributor to radio astronomy, to isolate genuine astronomical radio signals because she knew her equipment so thoroughly (L. Hooker 2004); and the sheer weight of building up laboratory skills layer by layer, from sterilization through tissue preparation to application of the experimental techniques that are essential to valid microbiology laboratory procedures.

26. Both values—ethical diversity for learning and the retention of local skilled cultures—are unavoidably in some tension with the idea of making learned improvements, which depend on impulses to generalization and unification. And it is a further issue whether those diversities should be taken as constituent features of any ethical or epistemic ideal toward which we should aim. We make no determination of these matters here as it would distract from our primary concern with the nature of ethical rationality and problem solving, but see also remarks on the inclusiveness of deliberation at section 7.5.

27. Design theorists are apt to emphasize the large cultural, social, and purposive differences between design and science and conclude that they generate and solve quite different problems, but this is a non sequitur because such differences are compatible with shared roles and structures; see Farrell and Hooker 2014 and section 4.3 below.

28. This will not in general be sufficient to determine the required construct, particularly when a very different construct ultimately proves to be required. Bridging

that under-determination is typically labeled creativity. The precise basis for that creativity remains unresolved. The three commonest options are these: (1) Inference. Constraints logically determine outcomes, e.g., induction. Ultimately, this proves futile. Thagard's notion of explanation as satisfaction of constraints (Thagard, 1988, Thagard and Verbeurgt 1998) and CzikCzentMihalyi's (1996a, 1996b, 1999) social constraints for a creative act leave the residual discovery process unspecified or random. (2) Emergence (e.g., adaptation in neural networks). Emergent processes do not specify their creative character and are in general not interpretable in agency functional terms. (3) Evolution (i.e., random search). Also ultimately futile: requisite searches are over ill-specified and infinite spaces, learning likely search heuristics too slow.

29. Of course they may also have smaller self-oriented cognitive purposes, e.g., to satisfy one's curiosity as to how this particular creature lives, irrespective of the value of the information to the larger community. But this is neither necessary nor sufficient to construct an epistemically powerful process.

30. Compare Kuhn's discussion (1977, chapter 13) of scientists who share a set of criteria for theory evaluation but disagree on the relative importance of these criteria in cases where these criteria yield conflicting judgments, and a similar discussion in Bjerring and Hooker 1979. On the mutual shaping of individual and institutional roles of relevance here, see, e.g., Vickers 1968, 1983.

31. Cf. "evolutionary drive" in Allen 2011. Thus the designs of real epistemic and ethical institutions become keys to their success. (On epistemic institutions as external nervous subsystem that organizes the body of science, see Hooker 1987a: 220–226, 309–315, Hooker 1995: 96–112.)

32. For some further elaboration see Hooker 1981, 2009, 2017a.

33. Cf. the criterion of progress in Hooker 1995, chapter 4. How this might apply in ethics is hinted in Hoffmaster and Hooker 2009—cf. Hooker 1994b and the suggested parallels between science and ethics in section 3.1—but awaits further development.

34. See Hooker 2002 for exposition.

35. On tensions, their ineliminability, and their productiveness, see Hooker 1987a, section 8.3.11.

36. For Piaget see, e.g., Piaget 1973, 1978 and the examination in Hooker 1995, chapter 5.

37. See Dorst and Cross 2001, Maher et al. 1996; cf. Cross 2006.

38. See also section 6.1, and section 9.1.5, ultimate two paragraphs. For the work on scientific research problems, see Farrell and Hooker 2007a, 2007b, 2009; for work on design, see Farrell and Hooker 2012, 2013, 2014, 2015; for work on the cognitive study of expertise, see, e.g., Christensen et al. 2015, 2016 and references; and for the

precursor work on self-directed anticipative learning, see Christensen and Hooker 2000a, 2000b, 2002. For other work on higher-order learning, see, e.g., "double-loop learning" (Argyris 1999) or "expanded rationality" (Hatchuel 2001). For more on the proposed universal problem-solving process, see Hooker 2017b.

Chapter 5

1. Subsequent references to this study in the text are given simply as page numbers.

2. Bluebond-Langner was tested in other ways: "It ... became apparent ... that the children asked me to watch TV to test the relationship. As one eight-year-old boy said to me after I watched TV for one solid hour without a word, 'All right. You're OK. What do you want to know?'" (1978: 247).

3. Bluebond-Langner adopts the account of mutual pretense in Glaser and Strauss 1965.

4. In this regard nonformal reason provides direction for critically examining the research on truth-telling to dying children that has appeared since Bluebond-Langner's study was published in 1978 and for designing future research. Unfortunately, the subsequent research is sparse, but see Kreicbergs et al. 2004 and Dunlop 2008.

5. Bluebond-Langner explains:The funeral is for the living, and so is the dying. The way we are permitted to die, and the way that we permit others to die, is to enable the living to continue the process of their lives. The dying have to be fitted into this scheme of the living, there is no separate way for them. Perhaps this is why, when we think of a "happy death," we think of dying suddenly or in our sleep. It is not just the physical pain of a long and lingering illness that we seek to avoid; for today we can almost guarantee people freedom from pain. No, it is the social pain, the pain that comes in the limbo phase, the pain of living with death, the pain that comes from committing an antisocial act, which we do not wish to endure. (1978: 233)

6. That has happened in part because of the repudiation by contemporary bioethics of paternalism and in part because this sense of autonomy is an expression of, and the moral support for, the legal doctrine of informed consent.

7. In addition to Zaner 1993, see, for example, Fins et al. 1999.

Chapter 6

1. One of us (BH) was once employed in that research.

2. On design process see, e.g., Cross 2006, Dorst and Cross 2001, Goel 1995, Lawson 2005, and Zeisel 2006.

3. See the account of specification in Richardson 1990.

4. Haslett makes a telling criticism of reflective equilibrium. He points out "the lack of any clear guidelines for making the many adjustment decisions that would have to be made in achieving a 'fit' among the various elements that go to make up a person's equilibrium," and he draws the conclusion for the limited rationality of the formal reason that dominates moral philosophy: "Notice that since [two conflicting options] (A) and (B) will both, by hypothesis, remove the contradiction and thus achieve coherence, coherence considerations alone are not enough to enable us to decide between (A) and (B). All coherence considerations enable us to decide is that one *or* the other must be chosen, they do not enable us to decide, definitively, *which* one. But then if we cannot decide between (A) and (B) by appeal to coherence, how *can* we rationally decide between them? Unless we are given a satisfactory answer to this question by proponents of reflective equilibrium methodology, and so far we have not been given a satisfactory answer, we shall, I think, be able to decide between (A) and (B) only arbitrarily, or else irrationally by appeal to prejudices, intuitions, or revelations." Haslett 1987: 310; emphasis in the original.

5. Winslow and Winslow recognize this feature of compromise in "a world of trade-offs": "the conflict will often be the inner, personal quandary that results when two or more of one's own deeply cherished moral values clash. Under these conditions, no decision will be without the accompanying sense of loss. There will be the residue of moral regret for the values lost in the alternatives not chosen" (Winslow and Winslow 1991: 313).

6. In this regard there is an obvious parallel between the philosophical rejection of moral compromise and the philosophical rejection of moral dilemmas. Both moral compromise and moral dilemmas are inconsistent with the logical consistency of moral principles. And given that, moral philosophy ignores or rejects the reality of what has been called moral "residue." A notable exception is Ruth Barcan Marcus, who recognizes the existence of residue and the value of residue-sustaining dilemmas: "although dilemmas are not settled without residue, the recognition of their reality has a dynamic force. It motivates us to arrange our lives and institutions with a view to avoiding such conflicts" (Marcus 1980: 121).

7. Benjamin's discussion of compromise in an intensive care unit is quite clear about this matter. He imagines two different outcomes that can result from a process of compromise that demonstrates mutual respect, personal honesty and insight, and flexibility. One is a conflict-removing specification that Benjamin regards as a compromise in only the "loose" sense. The other is a conflict-preserving compromise, thus a compromise in the "strict" sense. Of the former he declares that there is no compromise because there is no lingering conflict of obligations. The parties critically assess their initial principles and replace them with principles that have been rationally (in some unspecified sense of "rational") supplemented or amended and are regarded as superior to the principles that generated the conflict. Because the modified principles are mutually consistent, there is no conflict, hence no difference to be split, hence no compromise. For Benjamin only the latter outcome is a genu-

ine compromise because conflicts among obligations remain, and resolving them requires "splitting the difference." See Benjamin 1990 (chapter 3).

8. For the position developed in this section, and some further discussion, see Hoffmaster and Hooker 2017. See also Hoffmaster and Hooker 2013 for an arresting example.

9. Similarly, the discussion has found no inherent need for Benjamin's talk of people on both sides discovering "that their motivation is mixed, that their positions may be prompted by concern for themselves as well as concern for others." Such discoveries might induce willingness to compromise (because of a feeling that one's own position is weaker than at first thought), which is strategically useful, and they might indirectly induce humbleness in negotiation, which generally conduces to negotiability, but all of these useful consequences can be had directly simply as marks of ethical maturity. Talk of mixed motives as if it were a necessary condition for compromise is simply wrong.

10. For a conception of agency appropriate to such ideas, see Moreno and Mossio 2015, complemented by Skewes and Hooker 2009: 283–300, Christensen and Hooker 2002; cf. Hooker 2011b (section 4.1.1).

11. For further discussion, see sections II and III in Hooker 2010.

12. Of course, not all compromise expresses a conflict of *moral* obligations: a scientist typically cannot pursue all epistemic values simultaneously, for example, often not both width of explanatory power and explanatory precision in the same experimental design (an epistemic conflict); a business owner typically cannot pursue both efficiency and adaptability (a prudential conflict); and a person often cannot pursue both individuality, which requires social difference, and belonging, which requires social sameness (a mixed moral-prudential conflict). We confine ourselves to conflicts of moral obligations, although what we have to say extends directly to all the other kinds of conflict among valuable conditions.

13. And it matches nicely the conception of agency referenced at note 10 above and links to the development of expertise in problem solving at section 4.3.

14. Marcus (1980) makes the same point with respect to moral dilemmas.

15. In situations of finite deductive capacity, this is also true of the deduction-from-principles approach, for while complete deduction is formally specifiable, it is also unattainable, and there is no algorithm for deciding what finite subset of deductions it is best to choose to perform.

16. This is the moral correlate of the problem Hooker diagnoses in philosophy of science. See Hooker 1995: 306.

17. The mention of pertinent specific differences warns us that any simple formulaic approach, like maximizing QALYs, will often, perhaps mostly, be defeated by

the relevant variety among situations and participants. What approach could properly capture in a single formula the late discovery of writing, from mediocre but self-expressive to world significance, or the youthful joy of using hands again, from burping a baby to catching a pass in women's football? More subtly, how to capture the difference between experience simply enjoyed and experience that lays the groundwork for a later blooming of otherwise inaccessible life qualities, such as the captaining of a sporting team now that leads to heading a government agency later? Further, do deep enjoyment and transformative learning count as equal life qualities, or how to weigh them? There is a further complication, not much mentioned (and wholly suppressed in the economists' utility maximization structure), interrelations among values. For instance, it is a commonplace of our lives to introduce floors to valuing, for example that children are commonly forbidden to discuss sweet desserts before they have eaten their vegetables, that leisure spending ought to be curtailed if children go hungry. Deciding such matters is subtle and requires thorough exploration of the values involved, as they manifest in the specific situation, assessed through broad, critical appraisal. Total value can be considered only after these matters are settled and floors are satisfied. On occasion QALY models can be constructed in retrospect for such processes, but this cannot help with practical problem solving and, as the Winslows would say, doing that would eliminate moral perplexity in a way that is neither honest nor prudent. The approach to maximizing expected utility (a related model) taken by the women deciding whether to try to get pregnant after receiving genetic counseling (section 1.2) nicely illustrates their insights of this kind. In short, what rationality requires is thorough appraisal of each problem situation and its problem-solving process, in its idiosyncrasies and in any partial correspondences to wider principles, rules of thumb, etc. Typically there is nothing to rescue us from such complexities, no simple dominance—no supreme or overriding value, not even the value of life—and no satisfactory common currency like QALYs (another, constructed, form of dominance).

18. See, e.g., Hooker 2011c, section 6.

19. See Eliot 1948 on culture, and Hooker 2002 on Eliot in a regulatory systems setting.

20. A noteworthy exception is Nerlich 1989.

21. We are aware that lurking just beneath the surface here are further issues, such as the debate between classical economic rationality that calls for maximization of utility as the rational decision criterion and the satisficers who call for finding some sufficiently satisfactory increase in utility in lieu of maximization on the grounds that calculating maxima is often literally impossible for us finite creatures and even more often would cost more than finding it would deliver in benefit over a satisficing choice. We are also aware of the issue of whether to join the "rule utilitarians" in attempting to justify rules or principles pragmatically as convenient calculational-cost-saving devices for day-to-day ethical conduct or to join the "act utilitarians" in

trying to keep open the possibility of acting against any rule, no matter how entrenched, should the occasion demand. With respect to the former issue, we have declared ourselves firmly on the side of the satisficers (in fact earlier in this section), and in the case of the latter dispute, we are, therefore, firmly on the side of neither: pragmatically justified rules have their place but cannot be fundamental, but neither can maximizing utility be the fundamental justificatory criterion. Instead, a satisficing justification applies, accompanied with the imperative to look for (satisficingly feasible) improvements.

22. See note 2, especially Dorst and Cross 2001. For a development of the full problem-solving process, see Hooker 2017b.

Chapter 7

1. Rational deliberation may last from seconds to centuries. The time ultimately taken will typically be influenced by many internal and external factors, too numerous and situation-dependent to list here. Among the more important general relationships are that, roughly and all other things equal, increasing expertise is associated with increased quality of rationality and decreasing deliberative time required, and increasing quality of rationality is associated with longer deliberation time and increasing expertise. So, among other things, rapid judgments might reflect higher expertise rather than arationality. However, some (e.g., Haidt 2001) have thought that very short deliberative times signal arational, "intuitive" responses, that there are only post hoc rationalizations for these decisions, and that they constitute the vast majority of moral judgments. At best (and ignoring Haidt's tacit assumption of formal reason) this would curtail the scope of our position. But in fact we would argue that there is good reason to doubt that it is the entire story. Consider here the triumphs of the construction of modern science, engineering, law, and design, all primarily consequences of the accumulative learned application of rationality to develop deepening expertises. There is no serious prospect of reducing these to socialized rationalization (though, like fraud, some of this no doubt occurs); indeed, rational methodology is aimed at detecting and removing just such contaminants. Nor is there any plausible argument for just morality to be excluded from rationality. This would require an inexplicably fractured account of rational capacities (cf. Christensen and Hooker 2002), and a fracture that is inexplicably narrowly moral since there is ample evidence of our capacity to successfully engage in sophisticated social processes that require complex rational analyses (e.g., market economics, political treaty making). It is no surprise then that Haidt recognizes empirical support for a second, rational process of moral judgment formation that comes into play wherever difficulties in deliberation require and intervention is feasible. But the scope of such difficulties is very wide indeed, from snap character assessments to reconciling relativity with quantum mechanics. And, as noted, it is consistent with brief deliberations that they are the result of learned expertise

enabling swift more-rational-than-random responses. Moreover, it is common experience that rational processes are able to modify swift judgments, as when careful observation alters character assessment. And as history bears out, that includes developing increasing expertise in designing rational deliberation across all its domains of application. All told, we foresee a wide scope for application of this book's approach to nonformal rationality and moral deliberation and its particular focus on how to improve them both. And on all these same grounds we hold out hope for the cross-cultural applicability of this approach, while acknowledging that there can be subtle difficulties to such border crossings that should not be suppressed or underrated, sometimes including the nature and role of rationality itself. Like the preceding, we set these issues aside here in the interest of first developing the nonformal approach.

2. Creating narratives is a way of enhancing judgments, but in complex cases there often are competing narratives. In the most difficult cases, competing narratives conflict with learning processes themselves, both in their design and in the power relationships within those designs. Environmental conflicts, in particular, are vulnerable to these problems. Ottinger's (2013) insightful and illuminating ethnography of community-industry relations shows how citizen science, subsequent learning processes, and the pursuit of justice can be thwarted and co-opted, even with worst consequences, when learning processes and their accompanying narratives are institutionalized. We thank the anonymous reviewer who brought this acute feature of narratives to our attention.

3. For examples, see Brett and McCullough 1986 and Leikin 1989.

4. See similar cases in Zaner 1994, 2004.

5. See also Anspach 1993, chapter 3.

6. For a broader study of the impacts of institutional design on nurses and nursing, see Chambliss 1996. With respect to the former, Chambliss says: "The individual nurse and her setting are integrally joined; indeed, the individual and her setting are reciprocally, mutually defining" (1996: 10). With respect to the latter, he says: "The hospital itself is not a neutral setting but is actually the creator of many of the nurse's ethical difficulties. Ethical difficulties are not indicative of unintended flaws in the system but are instead expressions of that system in its most basic features" (11).

7. Beyond these considerations are other various contingent issues concerning who should be included or excluded in a deliberation, e.g., those who can present feminist and management insights and interests. Since these kinds of matters do not concern the rationality of deliberation per se, we do not pursue them here, but acknowledge their importance on many occasions. Some—e.g., feminist ones—are relevant so often and so rich in context that they have been dubbed "standpoints," and a special status for them has been suggested. (For a useful review, see Rolin 2016; for an extended discussion of feminist standpoint theory in science, see Hard-

ing 2015.) It is certainly not hard to see, for instance, how the quality of our social and medical knowledge and policies would be improved by effectively including feminist standpoints, and to this extent proper inclusion of it should be pursued. And recall that a robust sense of inclusion is intended, for instance, not only diversity of recruitment to existing medical procedures and treatments, but recruitment that could support the alteration of concepts, principles, and methods for understanding the issues and for activating policies that pertain to them (see section 7.1 above). Nonetheless, in deliberative design under nonformal reason they are no more than some considerations among others.

8. There is a substantial literature in these and related aspects of deliberation; see, e.g., Abelson et al. 2003, Cremaldi 2010, Ducet et al. 2001, Gauvin 2009, Gracia 2003, Misak 2008, Steinkamp and Gordijn 2003, Thornton 1994, Weidema et al. 2012, Weidema et al. 2013. Much of this literature would be sympathetic to our approach; however, it largely remains unintegrated (e.g., reviews of different "methods") and/or of a formulaic kind ("First set out the values implicated, second, ..."), which rarely picks up the richness of problem reformulation, method shift, self-transformation, and the like. Often it reveals continuing, tacit analytic assumptions; e.g., in an influential paper Gracia says, "Our moral decisions cannot be completely rational, due to the fact that they are influenced by feelings, values, beliefs, and so on, but they must be reasonable, wise and prudent." The reader should have some appreciation for how multiply problematic we would hold this sentence to be (while also avoiding affirmation of its opposite). A survey of 28 papers of this sort (Rigon et al. 2015) emphasizes the variety of designs used and of outcomes sought, from self-development to substantial communal agreement. There are websites devoted to assisting improved deliberation, and these often use checklist structuring of varying flexibility, from the strict "next do A" sort to those encouraging open-ended reflection on A and its alternatives. A checklist can provide a helpful initial framework, especially for a novice, but it may also serve to limit openness. And there is a literature on deliberation in a political context that often offers useful insight into ethical deliberation, e.g., Dillon 1994, Elster 1998, Mendes 2013, Van Belle et al. 2014.

9. In effect this is a parallel version of the social account of expertise as possession of tacit knowledge (see Priaulx et al. 2016). It accrues the positive benefits of that account, including provision of a basis for the roles of expertise in deliberation, in particular, in biomedical ethics (e.g., see Moreno 1991), but without having to restrict the knowledge to being tacit (though in varying degrees certain kinds may be, such as those requiring manipulative physical skills).

Chapter 8

1. For an illuminating account of how deep, often unrecognized, background assumptions and values create different approaches to disclosing a diagnosis of a terminal illness in Italy and the United States, see Gordon and Paci 1997.

2. Rapoport 1960: 11, quoting Luce and Raiffa 1957.

3. Indeed, one of us (CH) has championed adaptive resilience as the core of sustainability, where its natural synergy plays an important role. See Brinsmead and Hooker 2011.

4. Rapoport 1960: 234–242.

5. See, for example, Axelrod 1984, Campbell and Sowden 1985, Rapoport 1966, Richter 1985, and Sober 1992.

6. We could look for a formal way out. We could, for instance, attack the dominance principle. What dominance delivers is the guarantee that your outcome is the greatest it can be while being independent of what other players do, so this strategy excludes the cooperative potential of the situation. But choosing cooperation is a matter of nonformal judgment, not of logic; it is a question of how reliably accessible cooperation is given the context and the persons involved. However, this formal approach does highlight the fact that adopting dominance as a tool is itself also a matter of nonformal judgment, not simply a matter of logic or normative a priori truth. Furthermore, because it seeks superior performance only strategy-by-strategy, dominance has no global perspective; pursuing it can lead to making greater concessions to others than would be necessary. Brams (1976) discusses an example due to Howard (1971), a model of the 1962 Cuban missile crisis, where a dominant US (meta)strategy would, if played, have induced a Russian victory, whereas the cooperative equilibrium was actually reached. Its use, when it is used, will have to be based on more fundamental nonformal judgment as to its reasonableness. One alternative formal proposal derives from developing an extended "meta-game" whose strategies are concerned with conditional responses to the other player's choice of strategy (Howard 1971), where the Cpris/Cpris x/x position is the preferred stable outcome in this extended game. But this position must, for example, face the problem of the continuing incentive for individuals to defect and deceive pro tem. There is, for example, a formal regress argument from any last play (where defection is rational) to universal defection. Perhaps none of these positions is ultimately any easier to defend than the dismal alternative of arguing that z/z is the rational solution. There are strong grounds here for thinking that formal rationality cannot carry us forward.

7. There is sociological evidence from our individualist society that the difficulties cited in the text are real factors. See, for example, Bender and Mookherjee 1987, Glance and Huberman 1994a, 1994b, and Hardin 1982.

8. Communism functions equivalently to a traditional religious society in important respects, with its central call to place self second to the common good as a moral and cultural priority and its authoritative and authoritarian elders enforcing their versions of the sacred texts. And like traditional societies it resorts to a mix of reconversion and external coercion to control deviance.

9. There is a variant on external coercion in which the coercive disincentive to defect is provided directly by the other players, which we might designate MAP: mutually assured penalty. Its specific instance in nuclear deterrence is MAD: mutually assured destruction. Besides its demonstrated stability, it teaches us that in addition to the standard conditions for successful coercion, internal or external—that the coercer exists, can apply a sufficient penalty, and is a credible factor for the players—with MAP these conditions also have to survive the act of defection. Plausibly, many institutionalized social arrangements are of this kind. A weaker condition is MAM: mutually assured monitoring of compliance, with agreed cooperation but without backing by coercive penalties for defection beyond the threat not to cooperate in any relevant future repetitions. Schick (1992) argues that MAM is insufficient to prevent defection.

10. And similar conclusions would emerge from an examination of the Arrow paradox and from related problems in collective decision making. See, for example, Brams 1975 and 1976, McClennen 1990, Olsen 1965, Ordeshook 1986, Schick 1984, Sen 1982, Shubik 1964 and 1982, and Tversky and Kahneman 1986. Carens (1985) provides a nice study in the necessity of taking institutional structure into account when discussing rational social decision making, showing that one must institutionalize an ideal to properly understand it.

11. Authority and legitimacy, more attractive properties of law, are notoriously difficult to incorporate into the morality-free theories of legal positivists.

12. Fuller's *The Law in Quest of Itself* (1940) remains the most concise yet comprehensive account of the fundamental and intractable problems of legal positivism.

13. See, in particular, the essays in sections I and II of Fuller, *The Principles of Social Order* (1981).

14. Fuller's theoretical orientation also comports well with ours. In a critique of the pervasive "antinomy" between reason and fiat in case law, Fuller says that the many antinomies or paradoxes that are detected in the law "result from the attempt to treat a partial view of the law as if it were the whole view" (1946: 382). That is the legal manifestation of the totalizing impulse of philosophical theory construction. In the same vein Fuller commends Benjamin Cardozo for remaining "outside the arena of clashing doctrine" without becoming "merely eclectic" in his legal philosophy: His insight was too rich and varied, his method too flexible and too finely modulated to the task at hand, to make it possible for him to feel comfortable under the banner of any philosophic faction.

If his work lacked the sloganized kind of unity that would bring it under some familiar and inclusive rubric, this does not mean that it was not permeated by a deeper kind of coherence. The thread of connection in his philosophy is not found in theories or doctrines, but in a persistent effort to solve certain basic and recurring problems of the law. (Fuller 1946: 376).

Compare the quotation from Feyerabend for science in chapter 7, section 7.2.

15. For brevity, this discussion is taken from Fuller 1971: 334–337. For a more extended examination of this issue, see "Irrigation and Tyranny" in Fuller 1981.

16. See Lansing 2007, beautifully captured in the video "The Goddess and the Computer" (Lansing and Singer, Documentary Educational Resources, accessed 01/09/2017 at http://www.der.org/films/goddess-and-computer.html, and see further http://www.slansing.org/, accessed 01/09/2017), and Lansing 1996.

17. A water market, e.g., would introduce dependence on access to capital, tendencies to specialization and oligopoly, and so on, thus leading to very different landscape and communal configurations from those a water master may induce. Exactly this was the experience of water management in Bali (note 16). There is a substantial modern literature on management of irrigation systems, e.g., Ostrom 1992 (with chapter 4 on "Design Principles of Long-Enduring, Self-Organized Irrigation Systems") and Brown and Schmidt 2010. It encompasses all sides of this topic, and the reader is directed to it. We do not pursue it further here since our interest lies in more general issues concerning the ethics of evaluating and shaping institutional designs.

18. Fuller borrows the notion of a polycentric problem from M. Polanyi (1951).

19. At the time of their writing, Rittel and Webber were responding to the disappointment created by the failure of expectations that new systems approaches to problem solving would bring the social sciences within science and engineering and, more generally, to the failure of even grander claims that computational approaches to mind and artificial intelligence, engineering and formal management approaches to problem solving, and the like, would enable psychology, economics, and similar domains to be subsumed under the prevailing formal logical conception of scientific rationality. (Compare Cross 2006 here.) Although much has changed since then, their critical rejection of those traditional rationalizing ambitions remains widely supported throughout the literature on design and policy-making processes.

20. See Farrell and Hooker 2013. Backing the claim for nonformal rational methods in resolving wicked problems is a large literature on the practical design of effective problem solving, especially negotiation and processes drawn from business studies, conflict studies, relational and creative psychology, and so on that we do not have space to discuss here. Useful web pages (among hundreds) include ChangingMinds.org, "Discipline," http://changingminds.org/disciplines/disciplines.htm, Wikipedia, "Alternative Dispute Resolution," https://en.wikipedia.org/wiki/Alternative_dispute_resolution/. The idea of organizational design as an explicit design task is now many decades old. For typical entries from business and international studies, see, respectively, Passmore 1988 and Odell 2000.

Chapter 9

1. Blatchford, an experienced reporter who has spent many years covering the tragedies in criminal courts, has a sanguine view of Lachie's future: "I ... know that
Lachie will judge ... [his mother] kindly. The raw grief that was on his face in that
interview, his heart so plainly snapping—these things will never again show themselves that nakedly. He will rationalize, justify, even alter, those memories. He will
love his mum regardless. Children almost always do, no matter the size of the
betrayal" (2005: A3).

2. Styron adds, "For me, this transaction, with its imposition of guilt past bearing,
told more about the essential evil of Auschwitz than any of its most soulless physical
cruelties" (1999/1976: 601–602).

3. Calabresi and Bobbitt see affinities between the tragic choices they are examining
and Greek tragedy: "We have a prospect of insuperable moral difficulty, a nightmare
of justice in which the assertion of any right involves a further wrong, in which fate
is set against fate in an intolerable necessary sequence of violence." Arrowsmith
1965: 332. Quoted in Calabresi and Bobbitt 1978: 18.

4. See Sewall 1963: 120. Quoted in Calabresi and Bobbitt 1978: 19.

5. The information in this section is taken from "Concepts for Kidney Allocation,"
OPTN (2011).

6. See Formica 2014.

7. In discussing the allocation of artificial kidneys in England, Calabresi and Bobbitt
(1978) link efficiency and equality. They describe the method as "clinical judgment,
which could readily reduce itself to a sort of mechanistic, Newtonian efficiency-
determined egalitarianism" and explain that this approach would treat equally
those recipients who are equal with respect to exterior, observable therapy-related
criteria. The criteria are applied unswervingly and damn the implications for general
equality. Thus hemodialysis is allocated so as to achieve the highest rate of success,
given a limited number of kidneys available. (184)

   They go on to specify the forms of efficiency and equality operating here: This
mechanistic efficiency carries with it a certain kind of egalitarianism, by virtue of
what it excludes. That is, individual desires to live are, in the main, ignored; no
weighing of the relative value to society of one life or another is done either. Efficiency in this manner reduces itself to getting the most life-years out of the limited
number of machines. Everyone's desire to survive and live in treatment is assumed
to be equal. So is society's desire to have each one live. (185)

8. Calabresi and Bobbitt concentrate on three conceptions of equality—"absolute or
simple egalitarianism, formal egalitarianism designed to achieve a result which can
be termed efficient, and qualified egalitarianism which strives for efficiency but cor-

rects the efficient result to mitigate socioeconomic disadvantages"—but the second is amalgamated with efficiency, and the third is amalgamated with both efficiency and another conception of equality (1978: 178).

9. The charge of discriminating arbitrarily is disconcerting, however. Discrimination in its negative (social) connotation is not simply treating people differently, but treating people differently when there is no good reason for treating them differently. That kind of discrimination is intrinsically arbitrary. Categorization is not.

10. Calabresi and Bobbitt remark on the importance in the United States of "correcting second-order results before they reveal discriminations which offend a qualified egalitarianism, that is, that persons ought to be treated as equals if they are similar according to generalized efficiency criteria, but also if not treating them as equals displays a disfavored group in some prominent way. Here ... efficiency plays a significant role, but becomes suspect when it results in an obvious discrimination against a well-defined socioeconomic group" (1978: 186).

11. For a summary of this process, see Daniels 2000.

References

Abelson, J., Forest, P., Eyles, J., Smith, P., Martin, E., & Gauvin, F.-P. (2003). Deliberations about deliberative methods: Issues in the design and evaluation of public participation processes. *Social Science and Medicine*, 57(2), 239–251.

Allen, P. (2011). Complexity and management. In C. Hooker (Ed.), *Philosophy of complex systems*. Vol. 10 of Handbook of the philosophy of science. Amsterdam: North Holland/Elsevier.

Anspach, R. (1987). Prognostic conflict in life-and-death decisions: The organization as an ecology of knowledge. *Journal of Health and Social Behavior*, 28, 215–231.

Anspach, R. (1993). *Deciding who lives*. Berkeley, CA: University of California Press.

Anzai, Y., & Simon, H. (1979). The theory of learning by doing. *Psychological Review*, 86(2), 124–140.

Argyris, C. (1999). *On organizational learning*. Malden, MA: Blackwell Business.

Arpaly, N. (2003). *Unprincipled virtue: An inquiry into moral agency*. Oxford: Oxford University Press.

Arras, J. (1988). The severely demented, minimally functional patient. *Journal of the American Geriatrics Society*, 36, 938–944.

Arrowsmith, W. (1965). The criticism of Greek tragedy. In W. Corrigan (Ed.), *Tragedy: Vision and form*. San Francisco: Chandler Publishing.

Axelrod, R. (1984). *The evolution of co-operation*. New York: Basic.

Baier, K. (1958). *The moral point of view*. Ithaca, NY: Cornell University Press.

Beauchamp, T. (2003). The origins, goals, and core commitments of *The Belmont report* and *Principles of biomedical ethics*. In J. Walter & E. Klein (Eds.), *The story of bioethics*. Washington, DC: Georgetown University Press.

Beauchamp, T., & Childress, J. (1979). *Principles of biomedical ethics*. New York: Oxford University Press.

Beauchamp, T., & Childress, J. (2001). *Principles of biomedical ethics* (5th ed.). New York: Oxford University Press.

Bender, J., & Mookherjee, D. (1987). Institutional structure and the logic of ongoing collective action. *American Political Science Review*, 81, 129–154.

Benjamin, M. (1990). *Splitting the difference*. Lawrence: University Press of Kansas.

Bickhard, M. (1993). Representational content in humans and machines. *Journal of Experimental & Theoretical Artificial Intelligence*, 5, 285–333.

Bickhard, M. (2002). Critical principles: On the negative side of rationality. *New Ideas in Psychology*, 20, 1–34.

Bickhard, M. (2005). *The whole person: Toward a naturalism of persons*. In preparation.

Bjerring, A., & Hooker, C. (1979). Process and progress: The nature of systematic inquiry. In J. Barmark (Ed.), *Perspectives in metascience*. Lund: Berlings.

Blake, R., Ducasse, C., & Madden, E. H. (Eds.). (1960). *Theories of scientific method*. Seattle: University of Washington Press.

Blatchford, C. (2005). In the end, the children remain the bravest of all. *Globe and Mail* (Toronto). January 1, A3.

Bluebond-Langner, M. (1978). *The private worlds of dying children*. Princeton, NJ: Princeton University Press.

Bosk, C. (1992). *All God's mistakes*. Chicago: University of Chicago Press.

Brams, S. (1975). *Game theory and politics*. New York: Free Press.

Brams, S. (1976). *Paradoxes in politics*. New York: Free Press.

Breiman, L. (2001). Statistical modelling: The two cultures. *Statistical Science*, 16, 199–215.

Brett, A., & McCullough, L. (1986). When patients request specific interventions. *New England Journal of Medicine*, 315, 1347–1351.

Brinsmead, T., & Hooker, C. (2005). *Sustainabilities: A systematic framework and comparative analysis. Research Report 53*. Brisbane: Cooperative Research Centre for Coal in Sustainable Development.

Brinsmead, T., & Hooker, C. (2011). Complex systems dynamics and sustainability: Conception, method and policy. In C. Hooker (Ed.), *Philosophy of complex systems*. Vol. 10 of Handbook of the philosophy of science. Amsterdam: North Holland/Elsevier.

Brock, D. (1991). The ideal of shared decision making between physicians and patients. *Kennedy Institute Journal of Ethics*, 1, 28–47.

Brown, H. (1979). *Perception, theory, and commitment*. Chicago: University of Chicago Press.

Brown, H. (1988). *Rationality*. London: Routledge.

Brown, H. (1996). Psychology, naturalized epistemology, and rationality. In W. O'Donohue & R. Kitchener (Eds.), *The philosophy of psychology*. London: Sage.

Brown, H. (2006). *Decisions and the pursuit of knowledge*. Unpublished manuscript, Northern Illinois University, DeKalb, IL.

Brown, H. (2007). *Avoiding concepts*. Unpublished manuscript, Northern Illinois University. DeKalb, IL.

Brown, P., & Schmidt, J. (Eds.). (2010). *Water ethics: Foundational readings for students and professionals*. Washington, DC: Island Press.

Calabresi, G., & Bobbitt, P. (1978). *Tragic choices*. New York: W. W. Norton.

Campbell, N. (1957). *Foundations of science*. New York: Dover Publications.

Campbell, R., & Sowden, L. (Eds.). (1985). *Paradoxes of rationality and cooperation: Prisoner's dilemma and Newcomb's problem*. Vancouver: University of British Columbia Press.

Caplan, A. (1983). Can applied ethics be effective in health care and should it strive to be? *Ethics*, 93, 311–319.

Carens, J. (1985). Compensatory justice and social institutions. *Economics and Philosophy*, 1, 39–67.

Carnap, R. (1947). On the application of inductive logic. *Philosophy and Phenomenological Research*, 8, 133–148.

Carter, K., & Carter, B. (2005). *Childbed fever: A scientific biography of Ignaz Semmelweis*. Piscataway, NJ: Transaction Publishers.

Chambliss, D. (1996). *Beyond caring*. Chicago: University of Chicago Press.

Cherniak, C. (1986). *Minimal rationality*. Cambridge, MA: Bradford/MIT Press.

Christensen, W., Bicknell, K., McIlwain, D., & Sutton, J. (2015). The sense of agency and its role in strategic control for expert mountain bikers. *Psychology of Consciousness*, 23, 340–353.

Christensen, W., & Hooker, C. (2000a). Autonomy and the emergence of intelligence. *Communication and Cognition-Artificial Intelligence*, 17(3–4), 133–157.

Christensen, W., & Hooker, C. (2000b). An interactivist-constructivist approach to intelligence. *Philosophical Psychology*, 13, 5–45.

Christensen, W., & Hooker, C. (2001). Self-directed agents. *Canadian Journal of Philosophy* 31(suppl. 1), 18–52.

Christensen, W., Sutton, J., & McIlwain, D. (2016). Cognition in skilled action: Meshed control and the varieties of skill experience. *Mind & Language.* 31(1), 37–66.

Churchill, L. (1999). Are we professionals? *Daedalus*, 128(4), 253–274.

Cremaldi, A. (2010). *Deliberation in Aristotle's Ethics and the Hippocratic Corpus.* University of Pennsylvania. Publicly Accessible Penn Dissertations 192. Retrieved from http://repository.upenn.edu/edissertations/192/.

Cross, N. (2006). *Designerly ways of knowing.* Dordrecht: Springer.

Culyer, A. (2006). NICE's use of cost effectiveness as an exemplar of a deliberative process. *Health Economics, Policy, and Law*, 1, 299–318.

Czikszentmihalyi, M. (1996a). *Where is creativity?* New York: HarperCollins.

Czikszentmihalyi, M. (1996b). *Creativity: Flow and the psychology of discovery and invention.* New York: Harper Perennial.

Czikszentmihalyi, M. (1999). Implications of a systems perspective for the study of creativity. In R. Sternberg (Ed.), *Handbook of creativity.* New York: Cambridge University Press.

Daniels, N. (1996). *Justice and justification.* New York: Cambridge University Press.

Daniels, N. (2000). Accountability for reasonableness. *British Medical Journal*, 321(7272), 1300–1301.

Dewey, J., & Tufts, J. (1909). *Ethics.* New York: Henry Holt.

Diamond, J. (1998). *C.* London: Vermilion.

Dillon, J. (Ed.). (1994). *Deliberation in education and society.* Norwood, NJ: Ablex.

Dorst, K., & Cross, N. (2001). Creativity in the design process: Co-evolution of problem-solution. *Design Studies*, 22, 425–437.

Dorst, K., & Royakkers, L. (2006). The design analogy: A model for moral problem solving. *Design Studies*, 27(6), 633–656.

Doucet, H., Larouche, J.-M., & Melchin, K. R. (Eds.). (2001). *Ethical deliberation in multiprofessional health care teams. A project of the Saint Paul University, Centre for Techno-Ethics.* Ottawa: Ottawa University Press.

Dreyfus, H. (1991). *Being-in-the-world.* Cambridge, MA: MIT Press.

Dreyfus, H., & Dreyfus, S. (1986). *Mind over machine.* Oxford: Basil Blackwell.

Duhem, P. (1962). *The aim and structure of physical theory.* New York: Atheneum.

Dunlop, S. (2008). The dying child: Should we tell the truth? *Paediatric Nursing,* 20(6), 28–31.

Eliot, T. (1948). *Notes towards the definition of culture.* London: Faber.

Elster, J. (Ed.). (1998). *Deliberative democracy.* Cambridge: Cambridge University Press.

Emanuel, E., & Emanuel, L. (1992). Four models of the physician-patient relationship. *Journal of the American Medical Association,* 267, 2221–2226.

Farrell, R., & Hooker, C. (2007a). Applying self-directed anticipative learning to science I: Agency and the interactive exploration of possibility space in ape language research. *Perspectives on Science,* 15(1), 87–124.

Farrell, R., & Hooker, C. (2007b). Applying self-directed anticipative learning to science II: Learning how to learn across 'revolutions.' *Perspectives on Science,* 15(2), 222–255.

Farrell, R., & Hooker, C. (2009). Error, error-statistics and self-directed anticipative learning. *Foundations of Science,* 14(4), 249–271.

Farrell, R., & Hooker, C. (2012). The Simon-Kroes model of technical artifacts and the distinction between science and design. *Design Studies,* 33(5), 480–495.

Farrell, R., & Hooker, C. (2013). Design, science and wicked problems. *Design Studies,* 34(6), 681–705.

Farrell, R., & Hooker, C. (2014). Values and norms between design and science. *Design Issues,* 30(3), 29–38.

Farrell, R., & Hooker, C. (2015). Designing and sciencing: Reply to Galle and Kroes. *Design Studies,* 37(1), 1–11.

Feyerabend, P. (1974). Popper's objective knowledge. *Inquiry,* 17, 475–507.

Feyerabend, P. (1978). *Against method.* London: Verso.

Feyerabend, P. (1999). *Conquest of abundance.* Chicago: University of Chicago Press.

Fins, J., Bacchetta, M., & Miller, F. (1999). Clinical pragmatism: A method of moral problem solving. In G. McGee (Ed.), *Pragmatic bioethics* (1st ed.). Nashville, TN: Vanderbilt University Press.

Formica, R. (2014). New kidney allocation policy goes into effect Dec. 4. *American Kidney Fund Blog,* October 14. Available at http://www.kidneyfund.org/kidney-today/new-kidney-allocation-policy.html.

Frank, A. (2011). The philosopher as ethicist, the ethicist as storyteller. In O. Wiggins & A. Allen (Eds.), *Clinical ethics and the necessity of stories.* Dordrecht: Springer.

Fuller, L. (1940). *The law in quest of itself.* Chicago: Foundation Press.

Fuller, L. (1946). Reason and fiat in case law. *Harvard Law Review*, 59, 376–395.

Fuller, L. (1965). Irrigation and tyranny. *Stanford Law Review*, 17, 1021–1042.

Fuller, L. (1971). Mediation: Its forms and functions. *Southern California Law Review*, 44, 305–339.

Fuller, L. (1978). The forms and limits of adjudication. *Harvard Law Review*, 92, 353–409.

Fuller, L. (1981). *The principles of social order*. K. Winston (Ed.). Durham, NC: Duke University Press.

Galison, P. (1987). *How experiments end*. Chicago: University of Chicago Press.

Gallie, W. B. (1955–1956). Essentially contested concepts. *Proceedings of the Aristotelian Society*, 56, 167–198.

Gauvin, F.-P. (2009). What is a deliberative process? Canadian National Collaborating Centre for Healthy Public Policy. Retrieved from http://www.ncchpp.ca/docs/DeliberativeDoc1_EN_pdf.pdf.

Gilligan, C. (1982). *In a different voice*. Cambridge, MA: Harvard University Press.

Glance, N., & Huberman, B. (1994a). The dynamics of social dilemmas. *Scientific American*, 270(3), 76–81.

Glance, N., & Huberman, B. (1994b). Social dilemmas and fluid organizations. In K. Carley & M. Prietula (Eds.), *Computational organization theory*. Hillsdale, NJ: Lawrence Erlbaum Associates.

Glaser, B., & Strauss, A. (1965). *Awareness of dying: A study of social interaction*. Chicago: Aldine.

Goel, V. (1995). *Sketches of thought*. Cambridge, MA: MIT Press.

Gordon, D., & Paci, E. (1997). Disclosure practices and cultural narratives. *Social Science and Medicine*, 44, 1433–1452.

Gracia, D. (2003). Ethical case deliberation and decision making. *Medicine, Health Care, and Philosophy*, 6(3), 227–233.

Haidt, J. (2001). The emotional dog and its rational tail: A social intuitionist approach to moral judgment. *Psychological Review*, 108, 814–834.

Hardin, R. (1982). *Collective action*. Baltimore, MD: Johns Hopkins University Press.

Harding, S. (2015). *Objectivity and diversity: Another logic of scientific research*. Chicago: University of Chicago Press.

Harper, W., & Hooker, C. (Eds.). (1976). *Foundations of probability theory, statistical interference and statistical theories of science* (Vols. 1–3). Dordrecht: D. Reidel.

Haslett, D. (1987). What is wrong with reflective equilibria? *Philosophical Quarterly*, 37(148), 305–311.

Hatchuel, A. (2001). Towards design theory and expandable rationality. *Journal of Management and Governance*, 53–54, 260–271.

Heelan, P. (1988). Experiment and theory: Constitution and reality. *Journal of Philosophy*, 85, 515–524.

Heelan, P., & Schulkin, J. (1998). Hermeneutical philosophy and pragmatism: A philosophy of science. *Synthese*, 115, 269–302.

Hoffmaster, B. (1991). The theory and practice of applied ethics. *Dialogue*, 30(3), 213–234.

Hoffmaster, B. (Ed.). (2001). *Bioethics in social context*. Philadelphia: Temple University Press.

Hoffmaster, B. (2011). The rationality and morality of dying children. *Hastings Center Report*, 41(6), 30–42.

Hoffmaster, B., & Hooker, C. (2009). How experience confronts ethics. *Bioethics*, 23, 214–225.

Hoffmaster, B., & Hooker, C. (2013). Tragic choices and moral compromise: The ethics of allocating kidneys for transplantation. *Milbank Quarterly*, 91(3), 528–557.

Hoffmaster, B., & Hooker, C. (2017). The nature of moral compromise: Principles, values, and reason. *Social Theory and Practice*, 41, 55–78.

Hooker, C. (Ed.). (1973). *Contemporary research in the foundations and philosophy of quantum theory*. Dordrecht: Reidel.

Hooker, C. (1975). Global theories. *Philosophy of Science*, 42, 152–179. Reprinted as chapter 4 of Hooker 1987a.

Hooker, C. (Ed.). (1975–1979). *The logico-algebraic approach to quantum mechanics* (Vols. 1–2). Dordrecht: Reidel.

Hooker, C. (1976). Methodology and systematic philosophy. In R. Butts & J. Hintikka (Eds.), *Proceedings, 5th International Congress on Logic, Methodology and Philosophy of Science* (Vol. 3). Dordrecht: Reidel. Reprinted as chapter 5 of Hooker 1987a.

Hooker, C. (1981). Formalist rationality: The limitations of Popper's theory of reason. *Metaphilosophy*, 12, 248–266.

Hooker, C. (1982a). Scientific neutrality versus normative learning: The theoretician's and the politician's dilemma. In D. Oldroyd (Ed.), *Science and ethics*. Kensington: New South Wales University Press.

Hooker, C. (1982b). Understanding and control: An essay on the structural dynamics of human cognition. *Man-Environment Systems*, 12(4), 121–160. Reprinted as chapter 7 of Hooker 1987a.

Hooker, C. (1985). Surface dazzle, ghostly depths. In P. Churchland & C. Hooker (Eds.), *Images of science*. Chicago: University of Chicago Press.

Hooker, C. (1987a). *A realistic theory of science*. Albany, NY: State University of New York Press.

Hooker, C. (1987b). Evolutionary naturalist realism, circa 1985. In C. Hooker (Ed.), *A realistic theory of science*. Albany, NY: State University of New York Press.

Hooker, C. (1991). Between formalism and anarchism. In G. Munévar (Ed.), *Beyond reason: Essays on the philosophy of Paul Feyerabend*. Dordrecht: Kluwer Academic Publishers.

Hooker, C. (1994a). Idealisation, naturalism, and rationality. *Synthese*, 99(2), 181–231.

Hooker, C. (1994b). Value and system: Notes toward the definition of agriculture. *Journal of Agriculture and Environmental Ethics*, 7 (Special Supplementary Volume), 1–71.

Hooker, C. (1995). *Reason, regulation, and realism*. Albany, NY: State University of New York Press.

Hooker, C. (1998). Naturalistic normativity: Siegel's scepticism scuppered. *Studies in History and Philosophy of Science*, 29(4), 623–637.

Hooker, C. (2002). An integrating scaffold. In M. Wheeler, J. Ziman, & M. Boden (Eds.), *The evolution of cultural entities*. Oxford: Oxford University Press.

Hooker, C. (2009). Interaction and bio-cognitive order. *Synthese*, 166(3), 513–546.

Hooker, C. (2010). Rationality as effective organisation of interaction. *Axiomathes*, 21, 99–172.

Hooker, C. (Ed.). (2011a). *Philosophy of complex systems*. Vol. 10 of Handbook of the philosophy of science. Amsterdam: North Holland/Elsevier.

Hooker, C. (2011b). Introduction to philosophy of complex systems. Part A: towards framing a philosophy of complex systems. In Hooker 2011a.

Hooker, C. (2011c). Introduction to philosophy of complex systems. Part B: scientific paradigm + philosophy of science for complex systems: a first presentation c. 2009. In Hooker 2011a.

Hooker, C. (2017a). Re-modelling scientific change: Complex systems frames innovative problem solving. To appear in *Latu Sensu: revue de la Société de philosophie des sciences.* Submitted to *Proceedings of the European philosophy of science association conference 2016.*

Hooker, C. (2017b). A new model of innovative problem solving. *New ideas in psychology,* 47(December), 41–48.

Hooker, C., Penfold, B., & Evans, R. (1992). Control, connectionism and cognition: Toward a new regulatory paradigm. *British Journal for the Philosophy of Science,* 43, 517–536.

Hooker, G., & Hooker, C. (2017). Machine learning and the future of realism. *Spontaneous Generations,* 9(1). doi:10.4245/sponge.v9i1.27047

Hooker, L. (2004). *Irresistible forces.* Melbourne: Melbourne University Press.

Howard, N. (1971). *Paradoxes of rationality: Theory of metagames and political behaviour.* Cambridge, MA: MIT Press.

Howell, M., & Ford, P. (1980). *The true history of the elephant man.* New York: Penguin Books.

Ingelfinger, F. (1980). Arrogance. *New England Journal of Medicine,* 303, 1507–1511.

Kahneman, D., Slovic, P., & Tversky, A. (Eds.). (1982). *Judgment under uncertainty: Heuristics and biases.* Cambridge: Cambridge University Press.

Kaufman, S. (1997). Construction and practice of medical responsibility. *Culture, Medicine and Psychiatry,* 21, 1–26.

Keller, E. (1983). A feeling for the organism. In E. Keller (Ed.), *A feeling for the organism.* New York: W. H. Freeman.

Kreicbergs, U., Valdimarsdóttir, U., Onelöv, E., Henter, J.-I., & Steineck, G. (2004). Talking about death with children who have severe malignant disease. *New England Journal of Medicine,* 351, 1175–1186.

KTC (Kidney Transplantation Committee). (2012). Report to the Board of Directors, November 12–13. Available at http://optn.transplant.hrsa.gov/CommitteeReports/board_main_KidneyTransplantationCommittee_11_14_2012_11_33.pdf (accessed July 3, 2013).

Kuflik, A. (1979). Morality and compromise. In J. Pennock & J. Chapman (Eds.), *Compromise in ethics, law, and politics.* New York: New York University Press.

Kuhn, T. (1977). *The essential tension.* Chicago: University of Chicago Press.

Lakatos, I. (1976). *Proofs and refutations.* Cambridge: Cambridge University Press.

Lansing, J. (1996). *Perfect order: Recognizing complexity in Bali.* Princeton, NJ: Princeton University Press.

Lansing, J. (2007). *Priests and programmers: Technologies of power in the engineered landscape of Bali* (2nd rev. ed.). Princeton, NJ: Princeton University Press.

Lawson, B. (2005). *How designers think: The design process demystified* (4th ed.). Amsterdam: Elsevier.

Legg, C. (2005). The meaning of meaning-fallibilism. *Axiomathes*, 15(2), 293–318.

Leikin, S. (1989). When parents demand treatment. *Pediatric Annals*, 18, 266–268.

Levi, I. (1980). *The enterprise of knowledge*. Cambridge, MA: MIT Press.

Liao, S. (2011). Bias and reasoning: Haidt's theory of moral judgment. In T. Brooks (Ed.), *New waves in ethics*. London: Palgrave.

Lippman-Hand, A. (1982). Communication and decision making in genetic counseling. In T. Cohen, R. Goodman, & B. Bonné-Tamir (Eds.), *Human genetics: Proceedings of the Sixth International Congress of Human Genetics, Part B: Medical aspects*. New York: Alan R. Liss.

Lippman-Hand, A., & Fraser, F. (1979a). Genetic counseling: Provision and reception of information. *American Journal of Medical Genetics*, 3(2), 113–127.

Lippman-Hand, A., & Fraser, F. (1979b). Genetic counseling—the postcounseling period: I. Parents' perceptions of uncertainty. *American Journal of Medical Genetics*, 4(1), 51–71.

Lippman-Hand, A., & Fraser, F. (1979c). Genetic counseling—the postcounseling period: II. Making reproductive choices. *American Journal of Medical Genetics*, 4(1), 73–87.

Lippman-Hand, A., & Fraser, F. (1979d). Genetic counseling: Parents' responses to uncertainty. *Birth Defects Original Article Series*, 155C, 325–339.

Llewellyn, K. (1960). *The common law tradition: Deciding appeals*. Boston: Little, Brown.

Luce, R., & Raiffa, H. (1957). *Games and decisions: Introduction and critical survey*. New York: Wiley.

Maher, M., Poon, J., & Boulanger, S. (1996). Formalising design exploration as co-evolution: A combined gene approach. In G. Gero & F. Sudweeks (Eds.), *Advances in formal design methods for CAD*. London: Chapman and Hall.

Marcus, R. (1980). Moral dilemmas and consistency. *Journal of Philosophy*, 77, 121–136.

Margalit, A. (2004). *The ethics of memory*. Cambridge, MA: Harvard University Press.

Margalit, A. (2010). *On compromise and rotten compromises*. Princeton, NJ: Princeton University Press.

Marshall, P. (2001). A contextual approach to clinical ethics consultation. In Hoffmaster 2001.

Mattingly, C. (1998). In search of the good: Narrative reasoning in clinical practice. *Medical Anthropology Quarterly*, 12, 273–297.

McClennen, E. (1990). *Rationality and dynamic choice: Foundational explorations.* New York: Cambridge University Press.

McKeon, R. (Ed.). (1941). *The basic works of Aristotle.* New York: Random House.

Mendes, C. (2013). *Constitutional courts and deliberative democracy.* Published to Oxford Scholarship Online, April 2014.

Mill, J. (1843/2002). *A system of logic.* Honolulu, HI: University Press of the Pacific.

Mill, J., & Nagel, E. (Eds.). (1950). *Philosophy of scientific method.* New York: Hafner.

Miller, B. (1981). Autonomy and the refusal of lifesaving treatment. *Hastings Center Report,* 11(4), 22–28.

Misak, C. (2005). ICU psychosis and patient autonomy: Some thoughts from the inside. *Journal of Medicine and Philosophy,* 30, 411–430.

Misak, C. (2008). Experience, narrative, and ethical deliberation. *Ethics,* 118, 614–632.

Mitchell, C. (2014). Qualms of a believer in narrative ethics. In M. Montello (Ed.), *Narrative ethics: The role of stories in bioethics.* Hastings Center Report 44(1), S12–S15.

Montagu, A. (1971). *The elephant man: A study in human dignity.* New York: Outerbridge and Dienstfrey.

Moreno, A., & Mossio, M. (2015). *Biological autonomy: A philosophical and theoretical enquiry.* Dordrecht: Springer.

Moreno, J. (1991). Ethics consultation as moral engagement. *Bioethics,* 51, 44–56.

Nelson, H. (Ed.). (1997). *Stories and their limits: Narrative approaches to bioethics.* New York: Routledge.

Nerlich, G. (1989). *Values and valuing.* Oxford: Oxford University Press.

Odell, J. (2000). *Negotiating the world economy.* Ithaca, NY: Cornell University Press.

Oldroyd, D. (1986). *The arch of knowledge.* Kensington: New South Wales University Press.

Olsen, M. (1965). *The logic of collective action.* Cambridge, MA: Harvard University Press.

OPTN (Organ Procurement and Transplantation Network). (2011). Concepts for Kidney Allocation. Available at http://optn.transplant.hrsa.gov/SharedContent Documents/KidneyConceptDocument.PDF (accessed June 28, 2013).

Ordeshook, P. (1986). *Game theory and political theory*. Cambridge: Cambridge University Press.

Ostrom, E. (1992). *Crafting institutions for self-governing irrigation systems*. San Francisco: ICS Press.

Ottinger, G. (2013). *Refining expertise*. New York: New York University Press.

Passmore, W. A. (1988). *Designing effective organizations: The socio-technical systems perspective*. New York: Wiley.

Paulette, L. (1993). A choice for K'aila. *Humane Medicine*, 9(1), 13–17.

Piaget, J. (1973). *Main trends in psychology*. New York: Harper.

Piaget, J. (1978). *Behavior and evolution. Trans. D. Nicholson-Smith*. New York: Pantheon.

Pincoffs, E. (1986). *Quandaries and virtues*. Lawrence: University Press of Kansas.

Polanyi, M. (1951). *The logic of liberty*. Chicago: University of Chicago Press.

Popper, K. (1959). *The logic of scientific discovery*. London: Hutchinson.

Popper, K. (1972). *Conjectures and refutations* (4th ed.). London: Routledge and Kegan Paul.

Popper, K. (1979). *Objective knowledge* (rev. ed.). Oxford: Clarendon Press.

Priaulx, N., Weinel, M., & Wrigley, A. (2016). Rethinking moral expertise. *Health Care Analysis*, 24, 393–406.

Putnam, H. (1969). Is logic empirical? In R. Cohen (Ed.), *Proceedings of the Boston Colloquium in the Philosophy of Science 1966–1968*. New York: Humanities Press.

Putnam, H. (1982). Why reason can't be naturalized. *Synthese*, 52, 3–23.

Quill, T., Lo, B., & Brock, D. (1997). Palliative options of last resort. *Journal of the American Medical Association*, 278(23), 2099–2104.

Rapoport, A. (1960). *Fights, games, and debates*. Ann Arbor, MI: University of Michigan Press.

Rapoport, A. (1966). *Two person game theory*. Ann Arbor, MI: University of Michigan Press.

Rawls, J. (1951). Outline of a decision procedure for ethics. *Philosophical Review*, 60, 177–197.

Rawls, J. (1971). *A theory of justice.* Cambridge, MA: Belknap Press of Harvard University Press.

Rawls, J. (1974). The independence of moral theory. *Proceedings and addresses of the American Philosophical Association,* 48, 5–22.

Richardson, H. (1990). Specifying norms as a way to resolve concrete ethical problems. *Philosophy and Public Affairs,* 19(4), 279–310.

Richter, R. (1985). Rationality, group choice and expected utility. *Synthese,* 63, 203–232.

Rigon, C., Lourdes, E., & Vieira, M. (2015). Ethical deliberation in health: An integrative literature review. *Revista Bioética,* 23(1). doi:10.1590/1983-80422015231052.

Rittel, H., & Webber, M. (1973). Dilemmas in a general theory of planning. *Policy Sciences,* 4, 155–169.

Rolin, K. (2016). Values, standpoints, and scientific/intellectual movements. *Studies in History and Philosophy of Science,* 56, 11–19.

Ross, L., Parker, W., Veatch, R., Gentry, S., & Thistlethwaite, J., Jr. (2012). Equal opportunity supplemented by fair innings: Equity and efficiency in allocating deceased donor kidneys. *American Journal of Transplantation,* 12, 2115–2124.

Russell, B. (1912/2005). *The problems of philosophy.* London: Williams and Norgate.

Sack, K. (2012). In discarding of kidneys, system reveals its flaws. *New York Times,* September 19. http://www.nytimes.com/2012/09/20/health/transplant-experts-blame-allocation-system-for-discarding-kidneys.html.

Schick, F. (1984). *Having reasons: An essay on rationality and sociality.* Princeton, NJ: Princeton University Press.

Schick, F. (1992). Cooperation and contracts. *Economics and Philosophy,* 8, 209–229.

Schon, D. (1967). *Invention and the evolution of ideas.* London: Tavistock.

Sen, A. (1976). Rational fools: A critique of the behavioral foundations of economic theory. *Philosophy and Public Affairs,* 6, 317–344. Also in H. Harris (Ed.). (1979). *Scientific models of man.* Oxford: Oxford University Press.

Sen, A. (1982). *Choice, welfare and measurement.* Oxford: Blackwell.

Sewall, R. (1963). The tragic form. In L. Michel & R. Sewall (Eds.), *Tragedy: Modern essays in criticism.* Englewood Cliffs, NJ: Prentice-Hall.

Shapere, D. (1984). *Reason and the search for knowledge.* Dordrecht: Reidel.

Shi, Y. (2001). *The economics of scientific knowledge: A rational choice institutionalist theory of science.* Cheltenham, UK: Edward Elgar.

Shubik, M. (1964). *Game theory and related approaches to social behavior.* New York: John Wiley.

Shubik, M. (1982). *Game theory in the social sciences: Concepts and solutions.* Cambridge, MA: MIT Press.

Simon, H. (1947/1997). *Administrative behavior: A study of decision-making processes in administrative organization* (4th ed.). Glencoe, IL: Free Press.

Sinclair, W. (1909). *Semmelweis: His life and doctrine.* Manchester, UK: University Press.

Skewes, J., & Hooker, C. (2009). Bio-agency and the problem of action. *Biology and Philosophy*, 24(3), 283–300.

Sober, E. (1992). Stable cooperation in iterated prisoners' dilemmas. *Economics and Philosophy*, 8, 127–139.

Starzl, T., Hakala, T., Tzakis, A., Gordon, R., Steiber, A., Makowka, L., et al. (1987). A multifactorial system for equitable selection of cadaver kidney recipients. *Journal of the American Medical Association*, 257, 3073–3075.

Steinkamp, N., & Gordijn, B. (2003). Ethical case deliberation on the ward: A comparison of four methods. *Medicine, Health Care, and Philosophy*, 6(3), 235–246.

Stove, D. (1973). *Probability and Hume's inductive scepticism.* Oxford: Clarendon Press.

Styron, W. (1999). *Sophie's choice.* New York: Modern Library.

Thagard, P. (1988). *Computational philosophy of science.* Cambridge, MA: MIT Press.

Thagard, P., & Verbeurgt, K. (1998). Coherence as constraint satisfaction. *Cognitive Science*, 22, 1–24.

Thornton, B. (1994). The importance of process in ethical decision making. *Bioethics Forum*, 10(4), 41–45.

Tversky, A., & Kahneman, D. (1974). Judgment under uncertainty: Heuristics and biases. *Science, New Series*, 185(4157), 1124–1131.

Tversky, A., & Kahneman, D. (1986). Rational choice and the framing of decisions. *Journal of Business*, 58, 250–278.

Van Belle, H., Rutten, K., Gillaerts, P., Van De Mieroop, D., & Van Gorp, B. (Eds.). (2014). *Let's talk politics: New essays on deliberative rhetoric. Argumentation in Context* (Vol. 6). Amsterdam: John Benjamins Pub. Co.

Van Manen, M., & Kain, N. (2017). Developing morally sensitive policy in the NICU: Donation after circulatory determination of death. *Qualitative Report*, 22, 20–32.

Veatch, R. (1972). Models for ethical medicine in a revolutionary age. *Hastings Center Report*, 23, 5–7.

Veatch, R. (2000). *Transplantation ethics*. Washington, DC: Georgetown University Press.

Vickers, G. (1968). *Value systems and social process*. London: Penguin.

Vickers, G. (1983). *Human systems are different*. New York: Harper and Row.

Wasserstrom, R. (1961). *The judicial decision*. Stanford, CA: Stanford University Press.

Wedemeyer, P. (2010). Discussion of auditor judgment as the critical component in audit quality: A practitioner's perspective. *International Journal of Disclosure and Governance*, 7, 320–333.

Weidema, F., Molewijk, A., Widdershoven, G., & Abma, T. (2012). Enacting ethics: Bottom-up involvement in implementing moral case deliberation. *Health Care Analysis*, 20(1), 1–19.

Weidema, F., Molewijk, B., Kamsteeg, F., & Widdershoven, G. (2013). Aims and harvest of moral case deliberation. *Nursing Ethics*, 20(6), 617–631.

Whitbeck, C. (1996). Ethics as design. *Hastings Center Report*, 26(3), 9–16.

Whitbeck, C. (2011). *Ethics in engineering practice and research*. Cambridge: Cambridge University Press.

Winslow, B., & Winslow, G. (1991). Integrity and compromise in nursing ethics. *Journal of Medicine and Philosophy*, 16, 307–323.

Zaner, R. (1993). *Troubled voices*. Cleveland, OH: Pilgrim Press.

Zaner, R. (1994). Experience and moral life: A phenomenological approach to bioethics. In E. DuBose, R. Hamel, & L. O'Connell (Eds.), *A matter of principles?* Valley Forge, PA: Trinity Press International.

Zaner, R. (2004). *Conversations on the edge*. Washington, DC: Georgetown University Press.

Zeisel, J. (2005). *Inquiry by design: Environment/behavior/neuroscience in architecture, interiors, landscape, and planning* (rev. ed.). New York: W. W. Norton.

Index

Moral judgment (nonformal)
arrived at through moral deliberation
(*see* Deliberation, judgment and rationality; Designing deliberation)
rational when deliberation is rational (*see* Rationality (nonformal) and
moral judgment; Deliberation, judgment and rationality)

Narrative ethics, 172–179. *See also* Mattingly, Cheryl; Misak, Cheryl
Naturalism, 65–72
argument against, 69
argument for, 68
and finitude, 70–75
Normative-empirical relationship,
54–55, 68
and success, 69, 102
Nurse and doctor, differing clinical deliberation. See Intensive care unit
(ICU)

Occupational therapists, clinical deliberation of. *See* Mattingly, Cheryl

Polycentric problems, 214–217. *See also*
Fuller, Lon
Principles, moving beyond, 133–136
Public policy and tragic choices,
221–228
example: allocating kidneys for transplantation, 228–241

Rapoport, Anatol, on social interaction
modes. *See* Fights, games, debates
Rationality
and finitude, 70–75
formal, analytic (logical) model, critique of, 58–61
formal and nonformal reasoning, 4–5
naturalist foundation for nonformal,
non-analytic model (*see* Naturalism)

nonformal, non-analytic characterization of, 99–103
principles for construction of nonformal, non-analytic account, 20–21,
72–74
Western philosophical tradition, critique of, 61–65
Rationality (nonformal), and moral
judgment. *See also* Moral judgment
case examples (*see* Deliberation, judgment and rationality: case studies)
effective deliberation, 31–33
judgment rational when deliberation is
(nonformal) rational, 1–2, 144–146
moral effectiveness, 133–136
rational moral judgment, 39–41
rationality, from product to process,
41–45
skilled judgment, 75–78
Rationality (nonformal), resources for
creative construction, 90–95
observation, 79–86
reasoning (formal, nonformal), 86–90
skilled deliberative judgement, 75–78
synergy of resources, 97–99
systematic critical assessment, 95–97
Rational problem solving
in chapter 2 vignettes, 119–133
in childhood cancer case, 105–119,
and design, 100–103, 144–146
elements of in genetic counseling
case, 75–76,
universal model for, 100–101, 102–103
Reflective equilibrium, liberated, extended, enriched, 159–165

Science and ethics, deep similarities
between
historical development, 55–56
judgment, 53–54
normative learning, 54–55
problem solving, 53

Basic Bioethics

Arthur Caplan, editor

Books Acquired under the Editorship of Glenn McGee and Arthur Caplan

Peter A. Ubel, *Pricing Life: Why It's Time for Health Care Rationing*

Mark G. Kuczewski and Ronald Polansky, eds., *Bioethics: Ancient Themes in Contemporary Issues*

Suzanne Holland, Karen Lebacqz, and Laurie Zoloth, eds., *The Human Embryonic Stem Cell Debate: Science, Ethics, and Public Policy*

Gita Sen, Asha George, and Piroska Östlin, eds., *Engendering International Health: The Challenge of Equity*

Carolyn McLeod, *Self-Trust and Reproductive Autonomy*

Lenny Moss, *What Genes Can't Do*

Jonathan D. Moreno, ed., *In the Wake of Terror: Medicine and Morality in a Time of Crisis*

Glenn McGee, ed., *Pragmatic Bioethics*, 2d edition

Timothy F. Murphy, *Case Studies in Biomedical Research Ethics*

Mark A. Rothstein, ed., *Genetics and Life Insurance: Medical Underwriting and Social Policy*

Kenneth A. Richman, *Ethics and the Metaphysics of Medicine: Reflections on Health and Beneficence*

David Lazer, ed., *DNA and the Criminal Justice System: The Technology of Justice*

Harold W. Baillie and Timothy K. Casey, eds., *Is Human Nature Obsolete? Genetics, Bioengineering, and the Future of the Human Condition*

Robert H. Blank and Janna C. Merrick, eds., *End-of-Life Decision Making: A Cross-National Study*

Norman L. Cantor, *Making Medical Decisions for the Profoundly Mentally Disabled*

Margrit Shildrick and Roxanne Mykitiuk, eds., *Ethics of the Body: Post-Conventional Challenges*

Alfred I. Tauber, *Patient Autonomy and the Ethics of Responsibility*

David H. Brendel, *Healing Psychiatry: Bridging the Science/Humanism Divide*

Jonathan Baron, *Against Bioethics*

Michael L. Gross, *Bioethics and Armed Conflict: Moral Dilemmas of Medicine and War*

Karen F. Greif and Jon F. Merz, *Current Controversies in the Biological Sciences: Case Studies of Policy Challenges from New Technologies*

Deborah Blizzard, *Looking Within: A Sociocultural Examination of Fetoscopy*

Ronald Cole-Turner, ed., *Design and Destiny: Jewish and Christian Perspectives on Human Germline Modification*

Holly Fernandez Lynch, *Conflicts of Conscience in Health Care: An Institutional Compromise*

Mark A. Bedau and Emily C. Parke, eds., *The Ethics of Protocells: Moral and Social Implications of Creating Life in the Laboratory*

Jonathan D. Moreno and Sam Berger, eds., *Progress in Bioethics: Science, Policy, and Politics*

Eric Racine, *Pragmatic Neuroethics: Improving Understanding and Treatment of the Mind-Brain*

Martha J. Farah, ed., *Neuroethics: An Introduction with Readings*

Jeremy R. Garrett, ed., *The Ethics of Animal Research: Exploring the Controversy*

Books Acquired under the Editorship of Arthur Caplan

Sheila Jasanoff, ed., *Reframing Rights: Bioconstitutionalism in the Genetic Age*

Christine Overall, *Why Have Children? The Ethical Debate*

Yechiel Michael Barilan, *Human Dignity, Human Rights, and Responsibility: The New Language of Global Bioethics and Bio-Law*

Tom Koch, *Thieves of Virtue: When Bioethics Stole Medicine*

Timothy F. Murphy, *Ethics, Sexual Orientation, and Choices about Children*

Daniel Callahan, *In Search of the Good: A Life in Bioethics*

Robert Blank, *Intervention in the Brain: Politics, Policy, and Ethics*

Gregory E. Kaebnick and Thomas H. Murray, eds., *Synthetic Biology and Morality: Artificial Life and the Bounds of Nature*

Dominic A. Sisti, Arthur L. Caplan, and Hila Rimon-Greenspan, eds., *Applied Ethics in Mental Health Care: An Interdisciplinary Reader*

Barbara K. Redman, *Research Misconduct Policy in Biomedicine: Beyond the Bad-Apple Approach*

Russell Blackford, *Humanity Enhanced: Genetic Choice and the Challenge for Liberal Democracies*

Nicholas Agar, *Truly Human Enhancement: A Philosophical Defense of Limits*

Bruno Perreau, *The Politics of Adoption: Gender and the Making of French Citizenship*

Carl Schneider, *The Censor's Hand: The Misregulation of Human-Subject Research*

Lydia S. Dugdale, ed., *Dying in the Twenty-First Century: Toward a New Ethical Framework for the Art of Dying Well*